The Mathematics of Thermal Modeling

An Introduction to the
Theory of Laser Material Processing

John Michael Dowden

CRC Press
Taylor & Francis Group
Boca Raton London New York

CRC Press is an imprint of the
Taylor & Francis Group, an **informa** business

A CHAPMAN & HALL BOOK

CRC Press
Taylor & Francis Group
6000 Broken Sound Parkway NW, Suite 300
Boca Raton, FL 33487-2742

First issued in paperback 2019

ISBN-13: 978-1-58488-230-5 (hbk)
ISBN-13: 978-0-367-39731-9 (pbk)

Library of Congress Cataloging-in-Publication Data

Catalog record is available from the Library of Congress

**Visit the Taylor & Francis Web site at
http://www.taylorandfrancis.com**

**and the CRC Press Web site at
http://www.crcpress.com**

THE AUTHOR

John Dowden is a Professor of Applied Mathematics in the Department of Mathematics at Essex University in Colchester, England. He graduated from the University of Cambridge with a First Class Honors degree (B.A.) in Mathematics. In 1963 he enrolled as the first student of the University of Essex and obtained a Ph.D. in mathematical oceanography. In 1965 he joined the staff of the Mathematics Department as an Assistant Lecturer, where he has worked in a number of different positions, becoming a Professor in 1998.

Dr. Dowden is a Fellow of the Institute of Mathematics and its Applications. In addition, he is a member of a number of professional organizations including the Institute of Physics, the Laser Institute of America, and the European Institute for the Joining of Materials.

Dr. Dowden has been actively engaged in research into the construction of thermal models from physical principles in several fields. These have included industrially relevant problems such as welding, with a particular interest in innovative technology based on the use of lasers, as well as problems in laser medicine and volcanology. His work has involved him in collaboration with experts in the fields concerned from many different institutions and many different countries.

TABLE OF CONTENTS

PREFACE

The purpose of mathematical modeling is to find out what questions to ask next.

Bill Steen

The initial aim of this book is to describe the physical principles that need to be taken into account in modeling various aspects of laser material processing, and to formulate them in mathematical terms. The main purpose, however, is to show the way in which they can then be used to obtain insight into a number of processes by the construction of simple models whose underlying principles are easy to see. The immediate applications of the general principles considered here are welding, surface treatment, drilling, forming, and cutting, but the general principles have a very wide application. Many aspects of the theory are relatively independent of the material under study, which could be a simple metal, a complicated alloy (such as those used in the aerospace industry), concrete, or plastic. Although all the examples considered in this book have the common feature that the source of power is a laser, and the power of the laser is relatively high, the principles and methods are very general and apply to thermal modeling in many different fields, from microtechnology to laser medicine.

The emphasis is on the construction of *simple* models. This book is only concerned with those that can be solved to obtain worthwhile insights using analytical methods backed up with relatively simple numerical calculations that can be performed using standard general-purpose computer packages. Specific specialist software or the writing of long programs are not required; their use is a complementary skill to the kind presented here. Availability of general-purpose software capable of handling the routine numerical aspects of algebra and calculus and the evaluation of special functions is assumed, although it is not essential. It simplifies calculations very greatly, however. Computationally oriented modeling is a different kind of skill and is not addressed, but an understanding of the underlying physical ideas described should be helpful. Although the idea is not heavily emphasized, this book hopes to show how valuable it is to identify the relevant (as opposed to the irrelevant!) principles underlying the phenomena under investigation.

The book is intended primarily for engineers and material scientists at the Masters or first year Ph.D. level. It should also be of help to research workers coming to mathematical modeling of thermal processes in this area for the first time, whatever stage they have reached in their career development, or to mathematicians who have developed an interest in technological problems. The level of mathematical sophistication in modeling of this kind can vary a great deal. Some of the modeling described is relatively unsophisticated, some of it less so. The contents of the book are ordered by subject matter rather than mathematical sophistication, but as a rough guide, the degree of sophistication increases as each chapter progresses. The level of difficulty at the start of each new chapter is, however, considerably less than that at the end of the previous one. The exception is the first two sections of Chapter 2, which may be omitted at a first reading. The final section of that chapter summarizes the most important results required later on.

I would like to thank all those students, colleagues, and friends around the world who have encouraged me in the study of problems in modern technology. They have proved to contain a richness of interest for the mathematician that I had not expected. In particular, my appreciation goes to Phiroze Kapadia of Essex University and Bill Steen of Liverpool University. Phiroze has been a most stimulating colleague and collaborator with an extraordinary breadth of knowledge and depth of understanding of physics, and his determination and cheerfulness in the face of crippling illness is a continual inspiration. In the same way, Bill Steen's insight and enthusiasm for the problems of modern technology have been complementary driving forces. Between the two, I have been sure that, although I will be encouraged to see the heights of abstraction, I will never be allowed to lose sight of the practical goals of mathematical modeling.

Finally, I would like to thank my friends and family for their encouragement and forbearance during the writing of this book.

CHAPTER 1

THERMAL MODELING

1.1 INTRODUCTION

The availability of lasers as a source of thermal energy has led to their use in a variety of situations in many diverse fields, such as laser material processing and laser surgery.[1] They have the advantage that they can deliver known levels of power into small regions remarkably accurately; although a powerful laser is itself not small, its output can be delivered relatively simply in a manner that can be very accurately controlled. This feature makes lasers very suitable for automated processes. The fact that the absorbed power level can be estimated fairly accurately means that the effects of the thermal input, both the desired ones and the undesirable side-effects, can, in principle, be estimated relatively accurately. This is another advantage in work where precision is important, and is a principle reason why thermal modeling is valuable; but it has many aspects. It can be used, for example, to gain a fundamental understanding of the processes involved, or it can be used to demonstrate which effects are important and which are of lesser importance in a given context. It can be used to analyze situations where undesirable side-effects are discovered, or to save development time by means of elaborate computational models that have been shown to be capable of producing accurate numerical agreement with the results of earlier experiments.

It is a feature of mathematical modeling that has long been recognized, that many apparently very diverse phenomena can be modeled by essentially the same mathematics, and so it is here. The general approach to the modeling of laser surgery is not so very different to the use of lasers in the manufacturing industries. The power levels are dramatically different, but the underlying ideas are those of thermal conduction. There are, of course, other important differences,

---

[1] Ready, 1978; La Rocca, 1982; Mazumder, 1983; Steen, 1991; Hügel, 1994; Duley, 1996. Full publication details are given in the Bibliography.

but they can generally be handled in similar ways; experience in one area can be helpful in tackling problems in another. There is one point, however, that cannot be emphasized too much. When entering a new field, it is crucial for someone who wishes to apply their expertise to talk at length with experts in the field. Every field has its special interests and concerns as well as its own vocabulary. To interact with those in the area, a modeler must address problems of concern in a language the practitioners can understand, and must take account of the special features of the discipline concerned. A mathematician will be rewarded and often surprised by the diversity and interest of the mathematical problems that the approach will bring to light.

In the same way, people who come to mathematical modeling because of a prior interest in a particular area of application need a grasp of mathematics sufficient to do whatever job is of interest to them. In doing so they will learn how to use aspects of mathematics that they will later find useful in contexts other than the original one. The cross-fertilization that can occur in the process is one that can be immensely stimulating to all involved.[2]

The purpose of this book is not so much to catalog simple mathematical models as to try to give some idea by means of examples of why one should attempt to construct them in the first place. It is helpful to recognize what sort of considerations go into making something that will hopefully give greater insight, or to recognize what is the value and what are the dangers of approximations – and, indeed, to recognize approximations even when they come as accepted wisdom. There are a substantial number of sophisticated computer programs available that are dedicated to such things as the analysis of welding problems, or more generally in the fields of thermodynamics and computational fluid dynamics. Such programs can be of enormous use, but they need to be employed with understanding. Generally, they incorporate a great many aspects of the problems they address. For a proper understanding of the results the user needs to have some kind of feel, not only for the type of results to be expected of an experiment, but also the assumptions built into the program.

The situation with such tools is no different from that of the hand calculator for doing arithmetic: press the wrong buttons – or use the wrong algorithm – and you get the wrong answer. Users need to have

---

[2] See, for example, Tayler, 1986.

enough understanding to press the right buttons in the first place, and to recognize a nonsensical answer when they see it. Computer programs are automated versions of what are usually very sophisticated models; they differ from analytical models only in the way in which they obtain their answers. It is, however, often much easier to understand which mechanisms are important and which are not, and the kinds of results to be expected, from the study of simple examples. It is here that analytical models can so often be helpful.

Mathematical models can be developed at many different levels, and most people benefit by gaining experience with simple ones before trying more sophisticated examples. Not surprisingly, the mathematics needed vary enormously with the degree of sophistication of the model and the innate complexity of the problem to be studied. In problems of keyhole welding, for example, a familiarity with the cylindrical co-ordinate system is essential in a way that is not necessarily the case for other problems. The need for different levels of mathematical experience for different types of problems poses a difficulty for both the reader and the writer of a book such as this. The approach that has been adopted is to include some revision of the mathematics needed for the models studied in context. It is assumed that anyone who studies a given model has sufficient background, but may be unfamiliar or unpracticed in the use of the most advanced mathematics needed for the given example. The explanations given, therefore, are at the level appropriate to potential users of the given model, and are supplemented by references to standard mathematics texts. Such an approach results in an apparent unevenness in the background knowledge expected, but it is hoped that it results in a book that contains something of interest for people at a number of different levels of mathematical experience. It is perhaps worth pointing out that, although the material is ordered by subject matter rather than level of mathematical difficulty, it should be possible to read the book selectively, starting with the topics that involve the best-known mathematics. This should give confidence to the reader before moving on to those topics that require less familiar mathematics.

Logically, the equations of heat conduction, fluid dynamics, and the linear theory of thermoelasticity and the boundary conditions appropriate to them underlie all the material discussed. For that reason they are covered first, in Chapter 2. Any reasonably satisfactory account necessarily involves some discussion of the physical fundamentals and the way in which these can be converted to a suitable mathematical form. The process needs mathematics of a kind that tends to be little used outside contexts of this kind, and will appear to many to be unfamiliar. There are many ways of dealing with the problem, and the one adopted here is to try to present the arguments as economically as possible; the development of these ideas is not the main purpose of the book, necessary as it is. Some people may prefer to read other accounts, and a number of references are provided. Almost everyone will already be familiar with at least some parts of the theory. The results needed in the rest of the book are summarized in Section 2.3, and the two previous sections can be omitted at a first reading.

1.2 DIMENSIONS AND DIMENSIONLESS NUMBERS

   A feature that can at first be bewildering when people try to understand a mathematical model, or to construct one themselves, is the large number of quantities needed to describe a given problem. These can be genuine variables, such as the position in space of different points within a workpiece in which the temperature distribution is to be calculated, or properties of the material under consideration such as its density, thermal conductivity, etc. It is tempting at first to put in known numerical values for all those quantities that are just constants of the problem, just to reduce the number of symbols in use.

   Consider, for example, the following problem. A very thick sheet of steel has its underside cooled to a fixed temperature. Its surface stays at an average temperature of 15°C as a result of periodic heating with a nonzero mean part. It is known that the reflection coefficient is 50% and that the intensity of the incident power varies sinusoidally between 0 and 10 kW m^{-2} with a period of 1 minute. The problem is to find out what the highest temperature reached in the sheet is, at what time in the cycle this occurs, and at what temperature the underside must be maintained if the thickness of the sheet is 20 cm. (See Figure 1.1.)

   If z is the distance below the surface of the sheet and t is the time from the minimum value of the incident intensity, an approximate solution of this problem (with suitable values for the material properties of steel) is

$$T = 15 - 166.7z - 2.377\exp(-49.58z)\cos(0.1047t - 0.7854 - 49.58z).$$

$$(1.1)$$

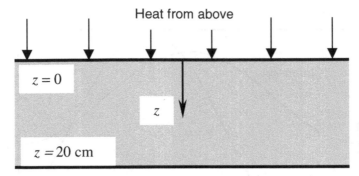

Figure 1.1. The plate heated from above and cooled from below.

Figure 1.2 shows isotherms as a function of z and t; it is clear that the greatest temperature occurs at the surface. Figure 1.3 shows the temperature at the surface and the incident intensity as a function of time. It is clear that the greatest temperature occurs $7\frac{1}{2}$ s after the greatest intensity. To answer the question as to what temperature the bottom surface should be kept at, it becomes clear that the answer is approximate, though an average temperature of about $-18\frac{1}{3}°C$ is required.

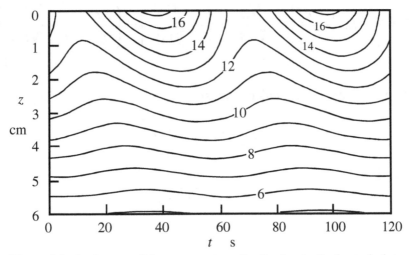

Figure 1.2. Isotherms of the temperature distribution in the heated plate in terms of depth z and time t over two cycles, with isotherms shown in °C.

Temperature scale   Intensity scale

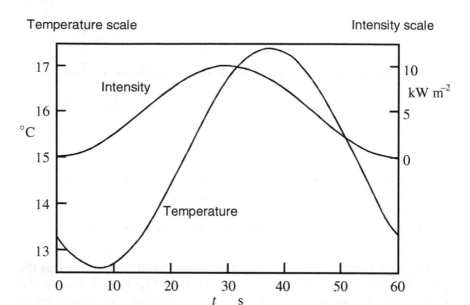

Figure 1.3. Temperature in °C at the surface and incident intensity as a function of time t over one cycle for the heated plate.

It is much more useful, however, to obtain[3] the solution in the form

$$T = T_0 - I_0 \frac{(1 - \mathcal{R})}{\lambda} \left\{ z + \sqrt{\frac{t_0 \kappa}{2\pi}} \, \exp\!\left(-z\sqrt{\frac{\pi}{t_0 \kappa}} \right) \cos\!\left(\frac{2\pi t}{t_0} - \frac{\pi}{4} - z\sqrt{\frac{\pi}{t_0 \kappa}} \right) \right\}$$

$$(1.2)$$

corresponding to the incident intensity

$$I = I_0 \left(1 - \cos\frac{2\pi t}{t_0} \right)$$

provided

$$h \gg \sqrt{t_0 \kappa / \pi} \ .$$

---

[3] The solution can be found in many ways; an outline of one method is given at the end of this section.

The symbols used are as in the notation in Section 2.3 with the addition of $\mathscr{R}$ for the reflection coefficient, T_0 for the mean surface temperature, I_0 for the average incident intensity (so its value is $10\,\mathrm{kW\,m}^{-2}$ for the numerical example above), and t_0 for the period.

What are the advantages? The following are just a few.

- At the most basic level, it is easier to check that you have done the algebra and calculus correctly. You only need to follow the symbols through, not redo all the arithmetic.
- There may well be simple intermediate checks – always valuable when expressions get complicated.
- The number of symbols can nearly always be reduced, sometimes very dramatically, and that is always an advantage if the algebra gets at all involved.
- It is possible to apply the same theory with different numbers, or in different contexts. In this example, suppose a different heating intensity or reflection coefficient were used, or suppose a different metal were chosen? The same theory applies, and it is just a matter of changing the numbers in the formula for the answer.
- It will usually be easier to see whether the approximations used in the construction of the model are worthwhile; there are inevitably some at one level or another, even though they may not be easily recognized.
- The nature of the answer obtained, when seen in a general form, may suggest ideas about problems that are mathematically similar but entirely different in terms of their origins in the real world.

In the above example it would not have been sensible to use the method employed to get the solution (1.2) if the plate had been only 5 mm thick, a result that is clear from the formula. The exponential multiplier of the periodic part of the solution would still be quite large when $z = 0.005$ m, a result that could have been predicted in advance from a consideration of the length scales in the problem, using dimensional analysis before any calculus had been used.

Dimensional analysis is based on the idea that, for example, a length is fundamentally different from a time; to speak of a rod as being 1 second long is simply not allowed. Although it is common usage in the U.S. to say that a town 50 miles distant is 1 hour away, there is an underlying assumption that the traveler's speed is 50 miles per hour or thereabouts. This means that distance d (1 mile) is related to time t (1 hour) by the relation $d = vt$, where v is the speed of travel (50 miles per hour). The result is a dimensionally valid relationship since the product of a velocity and a time is a distance. In scientific work, however, it is advisable to avoid verbal shortcuts of this kind.

It is possible to introduce separate symbols for the dimensions of dimensional quantities, but it is perfectly safe to use the symbols associated with the base units of the SI system. That is what will be done here, although the idea can be extended to entirely different contexts. The fundamental dimensional quantities needed are

Quantity	Symbol
length	m
time	s
mass	kg
temperature	K.

Because it is convenient to do so, and there is no risk of confusion, some of the main derived symbols of the system will be used, such as

$N = kg\ m\ s^{-2}$ for force, $J = N\ m = kg\ m^2\ s^{-2}$ for work,
$W = J\ s^{-1}$ for power, $Pa = N\ m^{-2} = kg\ m^{-1}\ s^{-2}$ for pressure.

The structure of the problem of the heated plate is really very simple, even though the answer does not appear so at first sight. The use of dimensionless variables can help in a number of ways. The following table lists the various quantities in the problem and their dimensions expressed in the base units of the SI system. Here, at least, there is no need to make a distinction between units and dimensions, although, strictly they are not the same thing.

Quantity	Units	Quantity	Units	Quantity	Units
T, T_0	K	t, t_0	s	z, h	m
λ	$W\ m^{-1}\ K^{-1}$	κ	$m^2\ s^{-1}$	I_0	$W\ m^{-2}$
$\mathcal{R}$	None				

The rules for dealing with dimensional quantities are

- all quantities combined by addition or subtraction must have the same units;
- the units of quantities formed by multiplication or division are obtained by treating the units as if they, too, can be multiplied or divided;
- standard functions should have dimensionless arguments.[4]

It is always safest to employ consistent sets of units; the SI base units are internationally accepted and are easy to use. All that is necessary is to keep track of powers of 10 – remember, for example, that 1 kW is actually 1000 W when doing arithmetic.

If h is the thickness of the plate, you can (if you really want to) add κ and h^2 / t_0, whose units are both $m^2\ s^{-1}$, but not I_0 and λ / h, whose units are $W\ m^{-2}$ and $W\ m^{-2}\ K^{-1}$, respectively.

These rules mean that at any stage in the construction of the formula, all terms in any given expression must have the same dimensions, and that includes the final formula itself. That provides a very simple check if symbols are used, as the dimensions of an expression obtained can be verified at any stage. If the check is satisfactory it will not, of course, guarantee that the expression is right; but if it fails, then you have made a mistake at some point and to continue without finding the mistake first would be a waste of time. If you substitute numbers at the start, such checks cannot be carried out.

Consider Equation (1.2). There is one term on the left, T, whose units are K, as are those of the first term on the right, T_0. The second term on the right is

$$- I_0 \frac{(1 - \mathcal{R})}{\lambda} z \quad \text{whose units are} \quad \left(W\ m^{-2} \right) \times \frac{1}{W\ m^{-1}\ K^{-1}} \times m = K$$

---

[4] The logarithm function, to whatever base, is an exception in a formal sense. It is, however, much safer to adhere to the rule in its case as well. If you define your own functions, it is also worth adhering to the rule in your definitions. Not to do so runs the risk of causing confusion as to the units in which quantities in your formulas should be measured, as there will probably be an implicit preferred system with the danger that an unwary user may use the wrong units. Your formula will look very different depending on the units employed.

and the third term on the right is

$$I_0 \frac{(1-\mathcal{R})}{\lambda} \sqrt{\frac{\kappa\tau_0}{2\pi}} \exp(...)\cos(...),\qquad(1.3)$$

whose units are

$$\left(\mathrm{W\,m^{-2}}\right)\times\frac{1}{\mathrm{W\,m^{-1}\,K^{-1}}}\times\sqrt{\frac{\mathrm{m^2 s^{-1}\times s}}{1}}\times1\times1=\mathrm{K}.$$

Remember that $\mathcal{R}$, π, and the standard functions exp and cos are dimensionless. All terms in the equation therefore have the dimensions of temperature; it only remains to check that the arguments of exp and cos are themselves dimensionless. Clearly the first and second terms in the argument of cos are, so it only remains to check the third argument of cos and the argument of exp. They are, in fact, the same:

$$-z\sqrt{\frac{\pi}{t_0\kappa}}\quad\text{whose units are}\quad \mathrm{m}\times\sqrt{\frac{1}{\mathrm{m^2 s^{-1}\times s}}}=1$$

as required.

Consider what happens if dimensionless variables are introduced in the given problem. Here, there is not very much choice about the way in which it can be done.

The natural time-scale is t_0, whose dimensions are s.

A little less obvious is the length scale. At first sight, h might seem the obvious choice, but if we are really interested in thick workpieces, it is not central and another choice has some advantages. From the fact that $-z\sqrt{\frac{\pi}{t_0\kappa}}$ is dimensionless, it is immediately clear that $\sqrt{t_0\kappa/\pi}$ has the dimensions of a length and would provide a satisfactory length scale. There is no need at all to include the factor involving π, but the resulting expressions will be a little bit simpler so there is perhaps an advantage.

T_0 has the dimensions of temperature, K, but is very much an irrelevance in this particular problem. It has no bearing on the answer other than to provide a baseline (and just to confuse things it was given in °C, as a more familiar unit in everyday use). It is therefore not a good choice. There is an alternative, as can be seen from (1.2); the dimensions of the second and third terms on the right are those of temperature, so we could use

$$I_0 \frac{(1-\mathcal{R})}{\lambda}\sqrt{\frac{t_0\kappa}{\pi}}$$

as the temperature scale. Again, the inclusion of the factor involving π and the absorption coefficient $(1-\mathcal{R})$ is entirely optional.

As shown in Chapter 2, the differential equation satisfied by T is

$$\frac{\partial T}{\partial t} = \kappa \frac{\partial^2 T}{\partial z^2};$$

if the dimensionless variables T', z', t' are introduced by writing

$$T = T_0 + I_0 \frac{(1-R)}{\lambda}\sqrt{\frac{t_0\kappa}{\pi}}\, T', \quad z = z'\sqrt{\frac{t_0\kappa}{\pi}}, \quad t = t_0 t'$$

the solution is expressible in the form

$$T' = -z' - \frac{1}{\sqrt{2}}\exp(-z')\cos\left(2\pi t' - \tfrac{1}{4}\pi - z'\right). \tag{1.4}$$

This is the solution of the problem given by

$$\frac{\partial T'}{\partial t'} = \pi \frac{\partial^2 T'}{\partial z'^2}, \quad -\frac{\partial T'}{\partial z'} = 1 - \cos 2\pi t' \text{ at } z'=0, \quad \int_{t'=0}^{1} T'(0,t')dt' = 0.$$

The problem as stated imposes the two surface boundary conditions explicitly, but a third condition is also given, that the underside of the plate is held at a constant (but unspecified) temperature. It is an important condition, as it leads to the rejection of any additional solutions of the problem, and has the form

$$T'(h',t') \text{ is independent of } t' \text{ at } h' = h\sqrt{\frac{\pi}{t_0\kappa}} >> 1.$$

Finally, a periodic solution is required so that there is no need to discuss transients. This requirement was not explicitly stated in the original description, but has been inferred from the absence of any reference to initial conditions. The inference might be incorrect! Whenever a mathematical model of some process is required, a major part of the process is working out what is actually of interest to the person making the request. It can sometimes be the most difficult part of the problem; there may well be all kinds of hidden assumptions that do not get mentioned since a person working every day in the field either regards them as obvious or is so used to them that they go unnoticed. The process of constructing the model will often throw a lot of light on hidden assumptions.

There are a number of points arising from the formulation of the problem in dimensionless terms. The problem was originally posed in a way that could arise quite naturally, but it does not correspond to the way in which one would go about finding the solution. To solve the problem, it is more natural to regard the temperature of the underside as given but not known, and then find the mean value of the surface. Since this is what is given, the temperature at which the underside must be held can then be inferred. The problem actually contains one *dimensionless number,* $h' = h\sqrt{\frac{\pi}{t_0\kappa}}$. It is, however, the only parameter in the formulation in dimensionless terms, demonstrating how economical the technique is in terms of notation. Dimensionless numbers can be very useful in deciding the kinds of approximations to be used, and relating problems from different contexts to one another. With a solution obtained for one value of the dimensionless number, a whole class of problems has been solved, not just one particular one. The results are then easy to transfer. Here we have, for example, solved the same type of problem for other metals, and perhaps even for a layer of ice on salt water (though one would have to examine the assumptions carefully in that case to be sure how relevant the solution was). Other advantages of the dimensionless form are that the individual stages in the derivation will be notationally much simpler, and that it may be much easier to spot general results.   In this example, it is now immediately obvious that the highest surface temperature occurs one-eighth of a period later than the maximum incident intensity. Perhaps this suggests an explanation of why it always seems hotter in the middle of the afternoon than at midday in hot weather (although, again, assumptions would have to be examined and the model would need to be modified somewhat to be satisfactory).   Surface cooling, for example, especially at night, is clearly important.

The phase lag is interesting and characteristic. Figure 1.4 shows the variation of temperature with depth relative to the mean value at time intervals of one-eighth of a cycle. The scales are those for Equation (1.1), but the pattern is similar for any solution of the form of (1.4). It will be seen how the amplitude of temperature about the mean value at a given depth oscillates with rapidly decreasing amplitude at greater depth, but the oscillations lag behind the values nearer the surface.

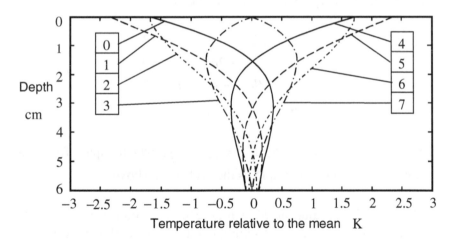

Figure 1.4. Variation of temperature with depth below the upper surface of the sheet relative to the mean temperature at that depth, at equal time intervals of one-eighth of a cycle. The curve labeled 0 is for the start of the cycle, and the one labeled 7 is seven-eighths of the way through; similarly for the other curves. The scales correspond to the example of Equation (1.1).

The scalings used to obtain (1.4) were suggested by the answer, but they could easily have been determined beforehand. The choice of t_0 is immediate, while the length scale can be deduced as follows. Rejecting T_0 and h as irrelevant, λ is the only remaining quantity involving K and so must be rejected; I_0 is then the only quantity left involving kg and cannot be used to construct a length scale. That leaves only κ (units, m^2 s^{-1}) and t_0 (units, s). The only combination of these with the dimensions of length is $\sqrt{t_0 \kappa}$ and dimensionless multiples of it. Similarly, the temperature scale must involve λ; that involves kg, so I_0 must also be involved. To remove time and length scales from it, κ and t_0 also have to be present. A very useful technique for finding the appropriate powers can be illustrated by this example. Suppose the necessary powers are a, b, c, and d, respectively. The units of $\lambda^a I_0{}^b \kappa^c t_0{}^d$ must be K, so

$$\left(\mathrm{kg\,m\,s}^{-3}\,\mathrm{K}^{-1}\right)^a \left(\mathrm{kg\,s}^{-3}\right)^b \left(\mathrm{m}^2\,\mathrm{s}^{-1}\right)^c (\mathrm{s})^d = \mathrm{K}.$$

This provides four equations for a, b, c, and d,

$$
\begin{aligned}
a \;+b \qquad\qquad\qquad &= 0 \\
a \qquad\; +2c \qquad\quad &= 0 \\
-3a \;-3b \;-c\; +d \;&= 0 \\
-a \qquad\qquad\qquad &= 1.
\end{aligned}
$$

Their solution is $a = -1$, $b = 1$, $c = \frac{1}{2}$, $d = \frac{1}{2}$, which, apart from the optional dimensionless multiple, is the scale used above.

The use of the value of h', which can seem important before the solution is obtained, can be a useful guide to the kind of approximations that might legitimately be used. In the problem as stated, the value of h' is nearly 10, which is reasonably large, especially as it appears in a negative exponential in the answer. If, however, the thickness had been 5 mm instead of 20 cm, h' would have been about $\frac{1}{4}$, which would have required a more complete solution; if it had been only 1 mm thick one might have been able to try a thin sheet approximation.

The purpose of the example was to illustrate how to use dimensional analysis, the use of dimensionless variables, and the value of dimensionless numbers. It is, however, a solution of the equation of heat conduction, so a brief indication will now be given as to how it can be obtained. The same kind of technique will be used in connection with pulsed heat sources in later chapters.

Perhaps the simplest way of finding the solution is to guess the form. Since there is a nonzero mean input, there has to be a mean vertical temperature gradient that must be linear to satisfy the time-independent conduction equation,

$$\frac{\partial^2 T}{\partial z^2} = 0.$$

To this, try adding a term whose form is $\mathrm{Re}\{Z(z)\exp(2\pi ti/t_0)\}$ where $i^2 = -1$. Z must then satisfy a second-order ordinary differential equation obtained by substitution of this form into the conduction equation and its derivative at the surface.[5] The value of the mean temperature gradient must satisfy the surface condition.

The same method can be used to find the solution when h' is of order 1. It is strongly recommended that the dimensionless formulation of the problem should be used when finding it, otherwise there is a danger of becoming lost in a bewildering forest of symbols.

1.3 TWO EXAMPLES

Thermal modeling can be applied to a great many industrial processes. As an illustration of its potential value, consider the cases of heat hardening of a metal workpiece and laser keyhole welding.[6]

CO_2 lasers have a high power density and, as a result, they are very suitable for some types of use in industry.[7] For example, it is possible to weld metal sheets of substantial thickness in a single pass. The energy

---

[5] Some familiarity with complex numbers is required with this approach; see Kreyszig, 1993, Ch.12. In particular, notice that the two square roots of i are $\pm\exp\frac{1}{4}\pi i$ i.e. $(1+i)/\sqrt{2}$ and $-(1+i)/\sqrt{2}$.

[6] Andrews, 1979.

[7] La Rocca, 1982; Mazumder, 1983.

can be placed very accurately, and so the thermal distortion of the work-piece is much less than by conventional means.[8] Similarly, they can be used for surface treatment of metals; lasers can produce surface hardening by martensitic transformation, for example, or overlay and diffusion coatings.[9] Because of the high power densities that can be achieved with lasers, a thin layer of the surface can be raised to temperatures well over 1000 °C very quickly so that very little thermal energy is absorbed per unit area. Heat distortion is therefore less, and sufficiently high rates of quenching can be achieved by conduction alone.

The technique is well established and empirical relations have been obtained between the process parameters.[10] Very general numerical models of the process have been constructed.[11] The problem was studied by analytical models from its inception.[12]

It is a characteristic of a laser beam that it can supply power at a very high intensity in a narrow beam. As a result, it is particularly suitable for penetration welding. Repeated passes are unnecessary if the laser beam is powerful enough, and since no material has to be removed to reach the bottom of the seam, filler rods are not needed.

There are other advantages to laser welding as well when contrasted with conventional welding techniques. For example, there is no physical contact with the workpiece so contamination problems are much less severe. Electron beam welding is in some respects rather similar, but normally has to be carried out in a vacuum. By contrast, laser welding can be carried out in the atmosphere. Furthermore, the heat affected zone is very small, unlike the situation with standard methods, though whether this is an advantage or a disadvantage will depend on the application. It is often possible to perform the welds in places that would normally be inaccessible, for example inside an enclosed vacuum tube. A number of materials that are otherwise difficult to weld can be processed. Titanium is not easy to weld with a TIG welder, for example. Probably the most serious disadvantages are the fact that the sizes of the pieces to be welded must be relatively small, and that the capital cost of the equipment needed is high.

---

[8] Megaw and Kaye, 1978.
[9] Similarly for electron beam welding, see e.g., Goldak et al., 1970.
[10] Steen and Courtney, 1979.
[11] For early examples see Henry et al., 1982, and Mazumder and Steen, 1980.
[12] Cline and Anthony, 1977; Ashby and Easterling, 1984.

Metal is vaporized when the laser is first directed at the material to be welded, while a hole forms through the material. Usually the hole penetrates to the far side of the workpiece but sometimes, either deliberately or accidentally, incomplete penetration occurs. This hole is referred to as a "keyhole;" when penetration is incomplete the keyhole is often said to be "blind." The workpiece is usually mounted on a moving table so that it can be moved relative to the laser. As it does so, the keyhole advances through the material. Welding will not occur if the power supplied is too high relative to the speed of translation. If the rate of absorption of energy by the workpiece is correct, however, only small amounts of material will be vaporized. The keyhole will then be surrounded by a region of liquid metal, known as the "weld pool." It is the resolidification of the molten metal that forms the weld. The hole itself is kept open by a combination of factors, but probably the most important are a combination of the pressure of the plasma that forms in the keyhole and the ablation pressure of the evaporating metal. Most of the material flows past the keyhole in the molten region and solidifies downstream of the laser, as shown schematically in Figure 1.5, though the details of the flow can be very complex.[13]

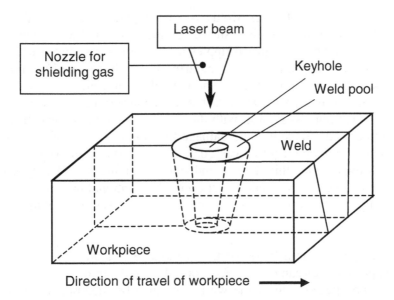

Figure 1.5. Configuration of the laser relative to the workpiece in laser keyhole welding.

---

[13] Matsunawa, 2000.

For electron beam welding, it has been demonstrated experimentally[14] that there can be a displacement of the elements of the material being welded when their final position in the workpiece is compared with their original position. In all probability the same is true of laser welding. The amount by which the material is shifted will depend on the relative sizes of the molten region and the keyhole.

Not a great deal of material is vaporized, but it is almost certainly very important in keeping the keyhole open, and perhaps also in determining the rate at which energy is transferred from the laser beam to the molten metal. As a result of the loss of material and differential stresses induced by the thermal cycle, the process may result in deformation of the workpiece, metallurgical changes with resulting changes in the properties of the material in and around the weld when compared to the surrounding material, or the production of residual stresses frozen into the material. Some of these factors could have important consequences for the strength and durability of the weld, and all of them are appropriate subjects for study using suitable physical and mathematical models.

A characteristic of laser welds is that liquid material is forced outward so that it forms a raised surface on the completed weld. On this there is superimposed solidified ripples in a pattern somewhat similar to herring bones in appearance, the regularity of which is sometimes taken as an indication of a good weld. The keyhole is at a pressure that will not normally be exactly the same as that of the atmosphere in which the welding operation takes place, with the result that metal vapor is forced out at either end. Since it could damage the optical system of the laser, it is usual to provide a jet of gas coaxially with the laser beam. The nozzle through which this flows is indicated in Figure 1.5. The type of gas used and the flow rate can affect optimal operational parameters and the quality of the weld. Two commonly used gases are helium and argon.

If the speed of translation of the workpiece is gradually increased various possibilities can occur. For example, the depth of the weld may decrease and penetration of the workpiece may be incomplete. When this happens, the depth achieved is often taken as the maximum depth of weld possible with this set of parameters. There is some ambiguity here since the phenomenon of "spiking" can occur in which there are sudden

---

[14] Basalaeva and Bashenko, 1977.

variations in the depth of penetration of the keyhole. These departures are usually of short duration, however. A slightly different concept, which is often taken to be the same, is the thickness of metal that can be reliably welded with a given set of parameters. Because the lower boundary of the workpiece affects the flow of thermal energy parallel to the laser beam, this depth is not the same as the maximum depth that can be achieved in a very thick workpiece. In the case of a blind keyhole, one that does not penetrate right through the workpiece, the actual depth of the weld is slightly greater since it is the depth of the molten region that defines the weld depth, not the length of the keyhole. In a great many models of keyhole welding all these concepts are considered to be the same, usually without discussion, for the simple reason that within experimental error they give effectively the same values. In transitional cases, however, for example between keyhole welding and conduction welding,[15] it is sometimes necessary to be aware of the distinctions.

A second transition phenomenon is that of "humping." If the translation speed is too high, the ripples on the surface of the weld pool may grow in amplitude to the point where they are so large that welding is no longer taking place in parts of the seam. Large solidified droplets may form on parts of the seam with no material in others; undercutting and voids can also occur. A different parameter zone is represented by the related technique of laser cutting, in which the molten material is removed entirely instead of being allowed to remain.

---

[15] Magee et al., 2000

CHAPTER 2

PHYSICAL PRINCIPLES

The first two sections of this chapter use more sophisticated mathematics than the rest of the book. The results obtained here, which are summarized in Section 2.3, are used in the rest of the book, but the derivations can be omitted at a first reading.

2.1 GOVERNING EQUATIONS

2.1.1 Conservation equations

Many ideas in the physical sciences can be considered as examples of conservation processes in which some quantity is neither created nor destroyed. The most familiar example of such an idea is perhaps the notion of *conservation of mass*. An extension of this idea occurs when a quantity is generally conserved but is also capable of being created or removed from the system. It may be worth remarking that these ideas often merge into one another and depend on the exact definitions of the quantities under consideration. An example might be the dissipation of mechanical energy into heat. In this instance mechanical energy is not conserved. Nevertheless, if the total energy is considered, consisting of both thermal and mechanical energy, and a sufficiently global view is taken, a conservation statement can be formulated. For reasons of this sort, all equations expressing ideas of this kind will be referred to as *conservation equations*. The convention will be followed even if, in a narrow sense, the principal quantity is only conserved in a global sense, or if it is more normally thought of as being generated by another type of quantity. An example of the latter is the way that force is related to the rate of change of momentum.

This chapter is concerned only with physical processes that are to be analyzed in terms of quantities that can be adequately described using continuous functions. Examples are the temperature distribution inside a solid body, or the body's density. Quantities that need to be studied in this connection are not necessarily *scalars* like the temperature and density examples just mentioned. The velocity at a point in a fluid in

21

motion, for example, is a *vector*, while the stress in an elastic solid is described in terms of the stress *tensor*. Most of the underlying equations considered in this book are mathematical statements of these ideas of conservation, generation, or destruction. They will be expressed in a form appropriate to representations of the physical quantities, in terms of mathematical functions that are sufficiently continuous for the techniques of differential and integral calculus to be used. Strictly speaking, physical quantities are not usually infinitely divisible in this way. The physical statements have to be made in terms of finite-sized regions, and a limit to zero size taken as a mathematical extrapolation. Such an approach is used because of the power and convenience of the techniques that are then available for the solution of problems of interest. A simple example is the *continuum approximation* of fluid mechanics. It is not possible to construct derivatives of the density of a fluid. The limiting process will pass through volumes in which the number of atoms is so small that it is not possible to talk meaningfully of a density. Mathematically, therefore, the density on these scales is a downward extrapolation of the values at bigger sizes. The same kind of idea applies to many other quantities.

All three cases, conservation, generation, and destruction, can, in these circumstances, be analyzed in the same way. It is possible to describe such ideas in terms of a simple partial differential equation. For simplicity all such equations will be referred to as *conservation equations*, even if they include generation or destruction terms. It is worth noting that generation and destruction can be included as a single effect differing only in its sign. A term sometimes used for equations of this type is *balance equations*.

Suppose there is some quantity, call it Q, that is defined at every point in some region of space D, and is associated with elements of the underlying material occupying D. The underlying material will often be the workpiece. Suppose

1. there is a flow $\boldsymbol{Q}$ of Q defined at each point of D so that the direction of flow is that of the unit vector $\hat{\boldsymbol{Q}}$ and the magnitude is $|\boldsymbol{Q}|$ units of Q per unit time;

2. there is a density k units of $\boldsymbol{Q}$ per unit volume defined at each point of D;

3. q units of Q are generated per unit time per unit volume of D. If there is destruction of Q, then adopt the convention that q is negative;

4. at any element of surface with unit outward normal **n**, there is a rate of generation of Q of **n**.$\mathcal{T}$ units of Q per unit area of S per unit time, where $\mathcal{T}$ will be assumed to be independent of the orientation of the surface element.

The first three of these assumptions are reasonably self-explanatory, but the fourth needs some explanation. In the simpler examples that will be considered here there is no such effect that needs to be investigated. Newton's second law of motion, however, shows that force generates momentum and can therefore be discussed in terms of a conservation condition; among the various types of force that can occur are pressure forces between material elements. These depend on the orientation of the surface elements in contact, and hence on **n**. In the simpler cases such as the dynamics of a Newtonian fluid, the remaining features of these forces can be described in a way that is compatible with Point 4, however. A similar situation occurs in the linear theory of elasticity.

Now consider any fixed volume V that will be taken to include its bounding surface S. It can be chosen in any way one wishes, provided that it lies in D. For the purposes of the following argument V, can be defined in a purely hypothetical way – it need not have physically observable boundaries. For the most part, such regions of material need exist only in the mind. For a great many quantities in physics there is a fundamental balance for the property Q in any such material volume V, given by the following statement.

The net rate of increase of Q in V is equal to the total rate of generation of Q inside V and on its surface S, less its net rate of outward flow across S.

See Figure 2.1.

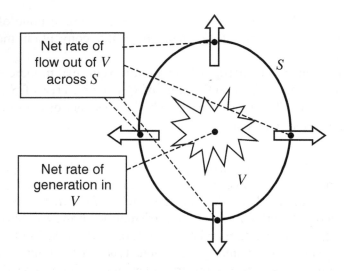

Figure 2.1. Contributions to the net rate of increase of Q in V.

The net rate of increase of Q in V is simply the rate of change of the total quantity of Q in V, which can be calculated as an integral of the density k. Consequently, it is equal to

$$\frac{d}{dt}\int_V k\, dV = \int_V \frac{\partial k}{\partial t} dV .$$

The rate of generation (or destruction, remembering the convention that q is measured positive for generation and negative for destruction) is given by

$$\int_V q\, dV + \int_S \mathbf{n}.\mathcal{J}\, dS .$$

Rather more difficult to calculate is the net rate of flow out of V across its surface S. Consider a small element in the surface of V whose area is dS and which is perpendicular to the unit vector $\mathbf{n}$. It is usual to adopt the convention that $\mathbf{n}$ points out of a closed surface such as S, as illustrated in Figure 2.2. Suppose, also, that the angle between the normal $\mathbf{n}$ and the flow vector $\mathbf{Q}$ in that region is θ. The only contribution to the flow out of V across dS is that part of $\mathbf{Q}$ that is perpendicular to dS; the part parallel to it does not contribute. The magnitude of the rate of loss is then $|\mathbf{Q}| \cos\theta\, dS$.

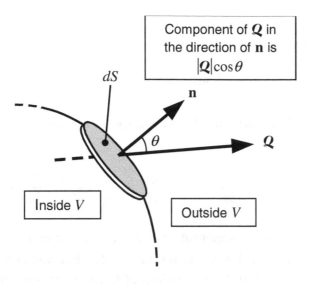

Figure 2.2. Flow out of *V* across S.

This is equal to **n.Q** *dS* , however, from the basic properties of the dot product of two vectors. In that case, there is a net flow rate

$$\int_S \mathbf{n.Q}\, dS \quad \text{units of Q}$$

out of *V*; the integral is a surface integral over *S*.[1] In consequence, the generation statement given above for the quantity can be expressed in mathematical form by the equation

$$\int_V \frac{\partial \boldsymbol{k}}{\partial t} dV = \int_V \boldsymbol{q}\, dV + \int_S \mathbf{n.}\boldsymbol{\mathcal{J}}\, dS - \int_S \mathbf{n.Q}\, dS\,, \qquad (2.1)$$

and it is true for any volume *V* inside *D*. This is the most general form of a conservation or generation law and is expressed in predominantly integral form. The arbitrariness inherent in the choice of *V* is a very important aspect of the equation, but it is much easier to investigate the properties of the system if this arbitrariness is removed. That can be done by converting the system into a partial differential equation by the following technique. Gauss's theorem, sometimes known as the divergence theorem, states that if a differentiable vector field **G** is

---

[1] Kreyszig, 1993, Ch.9.

defined in a volume V and on its surface S, which must be sufficiently smooth, then

$$\int_S \mathbf{n.G}\, dS = \int_V \nabla.\mathbf{G}\, dV .$$

The notation $\nabla.\mathbf{G}$ indicates the divergence[2] of the vector field $\mathbf{G}$ and is given by

$$\nabla.\mathbf{G} \equiv \operatorname{div}\mathbf{G} \equiv \frac{\partial g_1}{\partial x} + \frac{\partial g_2}{\partial y} + \frac{\partial g_3}{\partial z} \equiv \frac{\partial g_1}{\partial x_1} + \frac{\partial g_2}{\partial x_2} + \frac{\partial g_3}{\partial x_3} \equiv \frac{\partial g_i}{\partial x_i}$$

if the vector $\mathbf{G}$ has components (g_1, g_2, g_3) in terms of coordinates (x, y, z) or (x_1, x_2, x_3) of vector position $\mathbf{r}$. The last way of writing it is the form it takes when the conventions of Cartesian tensor notation[3] are employed and use is made of Einstein's summation convention. Cartesian tensors employ a notation in which all components of tensor or vector quantities are expressed in terms of a rectangular Cartesian co-ordinate system, and the components are distinguished by numerical subscripts. The summation convention implies summation from 1 to 3 in any multiplicative expression in which the same subscript indicated by a letter appears exactly twice, so that here, for example,

$$\frac{\partial g_i}{\partial x_i} \equiv \sum_{i=1}^{3} \frac{\partial g_i}{\partial x_i} \equiv \frac{\partial g_1}{\partial x_1} + \frac{\partial g_2}{\partial x_2} + \frac{\partial g_3}{\partial x_3}.$$

The use of a single dummy subscript usually means that the subscript can take any of the values from 1 to 3. Writing just g_i, for example, is a way of indicating all three components of the vector (g_1, g_2, g_3).

Gauss' theorem can therefore be used to convert the second and third terms on the right-hand side of the conservation statement (2.1) to volume integrals as well. In that form it now reads

$$\int_V \frac{\partial \mathbf{k}}{\partial t}\, dV = \int_V \mathbf{q}\, dV + \int_V \nabla.\boldsymbol{\sigma}\, dV - \int_V \nabla.\mathbf{Q}\, dV$$

---

[2] Kreyszig, 1993, Ch.8.
[3] Jeffreys, 1957.

$$= \int_V \left(\boldsymbol{q} + \nabla.\boldsymbol{\mathcal{J}} - \nabla.\boldsymbol{Q} \right) dV \, .$$

If it is now written as a single integral, we get the statement that

$$\int_V \left(\frac{\partial \boldsymbol{k}}{\partial t} + \nabla.\boldsymbol{Q} - \nabla.\boldsymbol{\mathcal{J}} - \boldsymbol{q} \right) dV = 0$$

for any volume V contained in D. The arbitrariness of V means that the integrand itself must vanish. For many people the result is intuitively obvious, but for others it is not. The way to see it formally is to assume that the integrand is continuous, divide the integral by the magnitude of the volume V, and allow the volume to tend to zero. The limit must be taken in such a way that V always contains an arbitrarily specified point of D and its overall dimensions also tend to zero.

The consequence of the argument is that

$$\frac{\partial \boldsymbol{k}}{\partial t} + \nabla.\boldsymbol{Q} = \boldsymbol{q} + \nabla.\boldsymbol{\mathcal{J}} \tag{2.2}$$

at each point in D, a generalized form of the conservation equation.

It is now possible to use all the methods of differential and integral calculus to investigate the properties of a system that possesses such a conservation law.

2.1.2 The equation of conservation of mass

Probably the simplest example of a conservation equation is that of the equation of conservation of mass in the form used in the study of the motion of a fluid when an Eulerian description of motion is employed. Suppose that the material under consideration has a velocity $\mathbf{u}$ (m s^{-1}) at a point whose position vector is $\mathbf{r}$, and that its density there is ρ (kg m^{-1}). In that case, $\boldsymbol{k}$ in Equation (2.2) is replaced by ρ, $\boldsymbol{q}$ by 0, $\boldsymbol{\mathcal{J}}$ by $\mathbf{0}$, and $\boldsymbol{Q}$ by $\rho\,\mathbf{u}$ since the flow of mass is entirely convective. The equation of conservation of mass is therefore

$$\frac{\partial \rho}{\partial t} + \nabla.(\rho\mathbf{u}) = 0 \, . \tag{2.3}$$

It can also be written

$$\frac{D\rho}{Dt} + \rho\nabla.\mathbf{u} = 0$$

where a widely-used abbreviation for the differential operator is

$$\frac{D}{Dt} \equiv \frac{\partial}{\partial t} + \mathbf{u}.\nabla .$$

An important special case is that of an incompressible fluid in which there is no change of density following the motion of the fluid. If a small volume of fluid is labeled, then the fluid bearing that label always has the same density wherever it is in the future, although its neighbors may have different densities. The idea can be expressed in mathematical form by considering a short interval of time δt; the idea of incompressibility means that

$$\rho(\mathbf{r} + \mathbf{u}\delta t, t + \delta t) = \rho(\mathbf{r}, t) + O(\delta t^2).$$

Since

$$\rho(\mathbf{r} + \mathbf{u}\delta t, t + \delta t) =$$

$$= \rho(\mathbf{r}, t) + \frac{\partial\rho}{\partial x_1} u_1 \delta t + \frac{\partial\rho}{\partial x_2} u_2 \delta t + \frac{\partial\rho}{\partial x_3} u_3 \delta t + \frac{\partial\rho}{\partial t} \delta t + O(\delta t^2)$$

$$= \rho(\mathbf{r}, t) + \frac{D\rho}{Dt} \delta t + O(\delta t^2)$$

from the first five terms in the Taylor Series expansion of ρ as a function of four variables, it follows that for an incompressible fluid

$$\frac{D\rho}{Dt} = 0. \tag{2.4}$$

Consequently, the equation of conservation of mass for an incompressible fluid is

$$\nabla.\mathbf{u} = 0. \tag{2.5}$$

In particular, Equation (2.5) applies to a fluid of constant density.

2.1.3 The equation of heat conduction

The quantity that is conserved, generated, or lost in the theory of heat conduction is thermal energy. The thermal energy density per unit mass of a material such as a solid or liquid workpiece, of density ρ and specific heat at constant pressure c_p, is $\rho \int c_p dT$ (J m^{-3}). This, therefore, is the quantity to be substituted for k in Equation (2.2). The rate of flow vector Q is given partly by Fourier's law as $-\lambda \nabla T$ (W m^{-2} s^{-1}), where λ is the coefficient of thermal conductivity, and partly as a convective component. The latter results from the motion of the underlying material. Heat is carried with the material as it moves with a flow rate $\rho \mathbf{u}$. In total, therefore, $Q = -\lambda \nabla T + \rho \mathbf{u} \int c_p dT$ here. From (2.2) the equation for heat conduction is then

$$\left\{ \frac{\partial \rho}{\partial t} + \nabla.(\rho \mathbf{u}) \right\} \int c_p dT + \rho c_p \left(\frac{\partial T}{\partial t} + \mathbf{u}.\nabla T \right) = \nabla.\lambda \nabla T + q$$

where q (W m^{-3} s^{-1}) is the rate of heating per unit volume and is the generation term q in (2.2). Once again $\mathcal{J} \equiv 0$. Because of the equation of conservation of mass (2.3), the equation simplifies to

$$\rho c_p \left(\frac{\partial T}{\partial t} + \mathbf{u}.\nabla T \right) = \nabla.\lambda \nabla T + q . \qquad (2.6)$$

In general, $\rho, c_p, q,$ and λ are not constants. They may depend on position in space, on time, and on T. In material processing problems it is usual to neglect the contribution to q that results from the working of viscous stresses, for example, as such contributions are usually very small compared to other heat sources.

A transformation that is sometimes useful when the thermal conductivity λ is a function of T only is to put

$$G = \int \lambda \, dT$$

since it simplifies the form of the equation slightly. Equation (2.6) then becomes

$$\frac{\partial G}{\partial t} + \mathbf{u}.\nabla G = \kappa \nabla^2 G + \kappa q \tag{2.7}$$

where κ is the *thermal diffusivity* $(\text{m}^2\,\text{s}^{-1})$ defined by

$$\kappa = \frac{\lambda}{\rho\, c_p}.$$

An alternative transformation that may be useful when ρ and c_p are functions of T only is to put

$$S = \int \rho\, c_p\; dT$$

so that (2.6) becomes

$$\frac{\partial S}{\partial t} + \mathbf{u}.\nabla S = \nabla.\kappa\nabla S + q. \tag{2.8}$$

The special case when $\rho, c_p,$ and k are all constants reduces the equation to

$$\frac{\partial T}{\partial t} + \mathbf{u}.\nabla T = \kappa \nabla^2 T + \frac{q}{\rho\, c_p}. \tag{2.9}$$

If, in addition, the entire medium is moving with a constant velocity U in the direction of the positive x axis,

$$\frac{\partial T}{\partial t} + U\frac{\partial T}{\partial x} = \kappa \nabla^2 T + \frac{q}{\rho\, c_p}. \tag{2.10}$$

If conditions are entirely steady, the time derivative is missing from the equation above. It is important to be clear that the absence of dependence on t only indicates that conditions in the frame of reference of the coordinate system are steady. Such a frame of reference is often convenient from the mathematical point of view, even though the properties under investigation depend on the thermal history of individual material elements. A case in point occurs when the metallurgical properties are modified. If the underlying material moves with a constant velocity giving a solution $T(x, y, z, t)$ of Equation (2.10), the temperature of material initially at (x, y, z) is given at a later time t' by

$$T(x, y, z, t; t') = T(x + U(t' - t), y, z, t').$$
(2.11)

The general case is more complicated; it is necessary to calculate the position of the material element at the time $t' > t$ from the known velocity field $\mathbf{u}(\mathbf{r}, t)$. If the position of the element is given by $\mathbf{r}'(\mathbf{r}, t; t')$, then its temperature at the later time is given by

$$T(\mathbf{r}'(\mathbf{r}, t; t'), t')$$
(2.12)

where $\mathbf{r}'$ is the solution of the partial differential equation

$$\frac{\partial \mathbf{r}'(\mathbf{r}, t; t')}{\partial t'} = \mathbf{u}(\mathbf{r}'(\mathbf{r}, t; t'), t')$$ which satisfies $\mathbf{r}'(\mathbf{r}, t; t) = \mathbf{r}$.

This equation states the condition that the velocity of the material element at time t' later than t calculated using $\mathbf{r}'$, which moves with the element, is the same as the velocity $\mathbf{u}$ it has when it occupies that position.

Equation (2.11) is easily derived from (2.12) by noticing that in Cartesian coordinates in the case of steady motion in the direction of the x-axis, $\mathbf{r} = (x, y, z)$ and $\mathbf{u} = (U, 0, 0)$. The solution of the equation for $\mathbf{r}'$ is therefore $x' = x + U(t' - t), y' = y, z' = z$. Substitution of these into (2.12) gives (2.11).

2.1.4 Dynamics of a continuous medium

The same kind of technique can be used to establish the equations of motion of a continuous medium from Newton's laws of motion. The first law is a conservation law of the most basic kind. The third law tells us how a composite body can be aggregated as a whole rather than having to consider it permanently as a collection of discrete parts. The fundamental quantity then becomes momentum. The derivation of

Equation (2.2) was explained as if Q were a scalar quantity, although that is not essential. It could be, for example, the ith component in a Cartesian tensor description of some more general tensor quantity. Here we will take it to be the ith component of momentum. In that case, take k to be ρu_i (kg m^{-3} s^{-1}) and Q to be the rate of flow of this quantity, $\rho u_i \mathbf{u}$. Newton's second law then shows that the volume rate of generation q of Q is given by ρF_i where $\mathbf{F}$ (N kg^{-1}) is the total body force per unit mass acting on an element of the medium.

The forces in a continuous medium can be of three types, internal forces between pairs of particles within V, body forces, and surface (or contact) forces. Examples of internal forces are mutual gravitational forces or electrical forces between elements within V. By Newton's third law these are equal and opposite and therefore sum to zero over the whole of V. They make no net contribution to the total force acting on material in V. Body forces are those such as gravitational forces caused by external material, and it is their total that is described by $\mathbf{F}$.

It is now necessary to analyze the surface forces[4] in order to identify the form of $\mathcal{J}$. They are the result of one body of material acting on an adjacent body, by processes such as frictional contact or pressure. It is usual to assume the Euler-Cauchy stress principle, which asserts that the effect of material outside V is equivalent to the existence of a force acting on each element dS of the surface of V. Its form is taken to be $\tau(\mathbf{r}, \mathbf{n}, t)dS$ where $\mathbf{n}$ is the outward unit normal at dS and τ (N m^{-2}) is the *surface traction* or *stress vector*. It can be shown[5] that there exists a *stress tensor* $\underset{=}{\mathbf{p}}$, whose components in a Cartesian coordinate system (x_1, x_2, x_3) are $\{p_{ij}\}$, and that τ is given in terms of it by

$$\tau_i = n_j p_{ji}.$$

The summation convention applies and all indices run from 1 to 3. A convenient alternative is the diadic notation $\mathbf{n}.\underset{=}{\mathbf{p}}$. The component of the surface forces in the direction of a unit vector $\mathbf{a}$ can be written in the form $n_j p_{ji} a_i$ or $\mathbf{n}.\underset{=}{\mathbf{p}}.\mathbf{a}$. Notice the way in which brackets and the order

---

[4] For a more detailed discussion see, for example, S.C. Hunter, 1983, 50-61.
[5] Hunter, 1983, 80-82.

of the components in the diadic notation indicate the way in which suffixes are contracted in the tensor notation.[6] In the notation of Cartesian tensors Equation (2.2) is

$$\frac{\partial k}{\partial t} + \frac{\partial \mathbf{Q}_j}{\partial x_j} = q + \frac{\partial \mathbf{\mathscr{T}}_j}{\partial x_j} , \qquad (2.13)$$

so when the conservation equation (2.2) is used to express the consequences of Newton's second law, $\mathscr{T}$ is replaced by p_{ij} , showing that

$$\frac{\partial (\rho u_i)}{\partial t} + \frac{\partial}{\partial x_j}\left(\rho u_i u_j\right) = \rho F_i + \frac{\partial p_{ji}}{\partial x_j} .$$

Use of the product rule for differentiation shows that

$$\left[\frac{\partial \rho}{\partial t} + \frac{\partial}{\partial x_j}\left(\rho u_j\right)\right]u_i + \left\{\rho\frac{\partial u_i}{\partial t} + \rho u_j\frac{\partial u_i}{\partial x_j} - \rho F_i - \frac{\partial p_{ji}}{\partial x_j}\right\} = 0 .$$

The equation of conservation of mass (2.3) shows that the first bracket is identically zero. The second term must therefore also be identically zero. Consequently, the equation governing the motion of a continuous medium using this form of description (usually known as an Eulerian description) takes the form

$$\rho\frac{\partial \mathbf{u}}{\partial t} + \rho\mathbf{u}.\nabla\mathbf{u} = \rho\mathbf{F} + \nabla.\underline{\underline{\mathbf{p}}} . \qquad (2.14)$$

The last term here can be written in the alternative notation of Cartesian tensors as

$$\left(\nabla.\underline{\underline{\mathbf{p}}}\right)_i = \frac{\partial p_{ji}}{\partial x_j} .$$

---

[6] So, for example, $n_j p_{ji} a_i$ is equal to $\mathbf{n}.\underline{\underline{\mathbf{p}}}.\mathbf{a} = (\underline{\underline{\mathbf{p}}}.\mathbf{a}).\mathbf{n} = \mathbf{a}.(\mathbf{n}.\underline{\underline{\mathbf{p}}})$, but not to $\mathbf{a}.\underline{\underline{\mathbf{p}}}.\mathbf{n}$ unless $\underline{\underline{\mathbf{p}}}$ is symmetric.

There is a special property possessed by $\underline{\underline{p}}$ that can be deduced in the same way. It is possible to take moments of Newton's second law of motion about any specified point **a**. The law then becomes a statement about the way in which moment of momentum (angular momentum) is generated by couples. Again, the account will be given using the notation of Cartesian tensors. Some further notation needs to be introduced, however. First, it is convenient to introduce the Kronecker delta δ_{ij} defined by

$$\delta_{ij} = \begin{cases} 1 \text{ if } i = j \\ 0 \text{ otherwise.} \end{cases}$$

Expressions involving the vector product have to be written in terms of the *alternating tensor* ε_{ijk}, which is defined so that

$$\varepsilon_{ijk} = \begin{cases} 1 \text{ if } ijk \text{ is a cyclic permutaion of } 1,2,3 \\ -1 \text{ if } ijk \text{ is an anticyclic permutation of } 1,2,3 \\ 0 \text{ otherwise.} \end{cases}$$

It then follows that, for example, $(\mathbf{b} \times \mathbf{c})_i = \varepsilon_{ijk} b_i c_j$ and $(\nabla \times \mathbf{c})_i = \varepsilon_{ijk} \dfrac{\partial c_k}{\partial x_j}$, results that are verified by enumerating all the components. A result of particular importance is the identity

$$\varepsilon_{ijk} \varepsilon_{lmk} = \delta_{il} \delta_{jm} - \delta_{im} \delta_{jl}, \tag{2.15}$$

which is the tensor form of the standard vector identity

$$\mathbf{b} \times (\mathbf{c} \times \mathbf{d}) = (\mathbf{b}.\mathbf{d})\mathbf{c} - (\mathbf{b}.\mathbf{c})\mathbf{d}.$$

Both the tensor and the vector versions of the identity can be proved by enumerating components.

In that case, the underlying property Q is the ith component in a Cartesian frame of reference of moment of momentum about **a** with its density k given by $[\rho(\mathbf{r}-\mathbf{a})\times\mathbf{u}]_i = \varepsilon_{ijk}\rho(x_j - a_j)u_k$. The flux $\mathbf{Q}$ is then $\rho[(\mathbf{r}-\mathbf{a})\times\mathbf{u}]_i\,\mathbf{u}$, and is therefore a second-order tensor with components $Q_{ij} = \varepsilon_{ijk}\rho(x_j - a_j)u_k u_j$. The volume generation term q is simply $[(\mathbf{r}-\mathbf{a})\times\rho\mathbf{F}]_i = \varepsilon_{ijk}(x_j - a_j)\rho F_k$. To find the form of $\mathcal{T}$, consider $\mathbf{n}.\mathcal{T}$; this is $[(\mathbf{r}-\mathbf{a})\times(\mathbf{n.p})]_i = \varepsilon_{ilk}(x_l - a_l)n_j p_{jk}$, so it, too, is a second-order tensor with components $\mathcal{T}_{ij} = \varepsilon_{ilk}(x_l - a_l)p_{jk}$. (The order and bracketing of the diadic expression is important. It should be remembered that $\mathbf{n.p}$ is a vector whose i^{th} component is $n_j p_{ji}$.)

The conservation condition, Equation (2.2) in its tensor form (2.13), then shows that

$$\frac{\partial}{\partial t}\left\{\varepsilon_{ijk}\rho(x_j - a_j)u_k\right\} + \frac{\partial}{\partial x_j}\left\{\varepsilon_{ijk}\rho(x_j - a_j)u_k u_j\right\} =$$

$$\varepsilon_{ijk}(x_j - a_j)\rho F_k + \frac{\partial}{\partial x_j}\left\{\varepsilon_{ilk}(x_l - a_l)p_{jk}\right\}$$

Use the equation of conservation of mass, (2.3), and subtract the vector product of (2.13) with $\mathbf{r}-\mathbf{a}$. These operations show that

$$\rho u_l \frac{\partial}{\partial x_l}\left\{\varepsilon_{ijk}(x_j - a_j)u_k\right\} - \varepsilon_{ijk}\rho(x_j - a_j)u_l \frac{\partial u_k}{\partial x_l}$$

$$= \frac{\partial}{\partial x_j}\left\{\varepsilon_{ilk}(x_l - a_l)p_{jk}\right\} - \varepsilon_{ilk}(x_l - a_l)\frac{\partial p_{jk}}{\partial x_j}$$

$$= \varepsilon_{ilk}\delta_{jl}p_{jk} = \varepsilon_{ijk}p_{jk}.$$

The result that $\dfrac{\partial x_l}{\partial x_j} = \delta_{jl}$ has been used. For the same reason, the left-hand side is equal to $\rho\varepsilon_{ijk}u_j u_k$, which is zero since any given pair of different values for the subscripts j and k appear twice, once in cyclic order and once in anticyclic order. Their sum is therefore zero. In consequence, this equation shows that

$$\varepsilon_{ijk} P_{jk} = 0.$$

Multiply this equation by ε_{imn} and sum over i. Identity (2.15) shows that $p_{mn} - p_{nm} = 0$ or, equivalently, that

$$p_{ij} = p_{ji} \text{ for all } i, j = 1...3. \qquad (2.16)$$

In other words, the stress tensor is necessarily symmetric.

It is sometimes useful to consider the orientation of a surface in the material that corresponds to a maximum or minimum value of the normal component of the traction on it. Suppose the surface has a unit normal $\mathbf{n}$ with components n_i. What is required is the stationary value of $\tau_i n_i = n_i p_{ij} n_j$, as the orientation of $\mathbf{n}$ is varied. Such a variation has to be performed subject to the constraint that $\mathbf{n}^2 = 1$. The easiest way to find the values of $\mathbf{n}$ is to use the method of Lagrange multipiers.[7] The technique is to vary

$$L = p_{ij} n_i n_j - p\left(n_i^2 - 1\right)$$

with respect to all three components of $\mathbf{n}$ and the Lagrange multiplier, p. The use of the standard techniques for finding stationary values using calculus shows that

$$\frac{\partial L}{\partial n_i} = 2\left(p_{ij} n_j - p n_i\right) = 0, \quad i = 1...3 \text{ and } n_i^2 = 1;$$

in other words, there are in general three such directions $\mathbf{n}$, which are the eigenvectors[8] of the matrix corresponding to $\underline{\underline{\mathbf{p}}}$ with eigenvalues p.

The values of p are called the principal stresses. A positive value of p corresponds to a tensile stress, and a negative value to a compressive stress on the element of surface whose normal is $\mathbf{n}$. The tractions in these special cases are, in fact, normal to the surface element since they are proportional to $\mathbf{n}$.

---

[7] Swokowski et al., 1994, Ch.12.
[8] Kreyszig, 1993, Ch.7.

2.1.5 Euler's Equation for an ideal fluid.

An ideal fluid is one in which all surface stresses are perpendicular to the surface, so that

$$\mathbf{n}.\mathbf{p} = -p\mathbf{n}.$$

In consequence, $\underline{\mathbf{p}}$ must be $-p\underline{\underline{\mathbf{I}}}$ where $\underline{\underline{\mathbf{I}}}$ is the unit diadic, or equivalently

$$p_{ij} = -p\delta_{ij}$$

where δ_{ij} is the Kronecker delta, equal to 1 when $i = j$ and zero. Otherwise, p is the pressure and is positive when the forces are compressive in character. Equation (2.2) therefore becomes

$$\rho\frac{\partial \mathbf{u}}{\partial t} + \rho\mathbf{u}.\nabla\mathbf{u} = \rho\mathbf{F} - \nabla p. \qquad (2.17)$$

This is known as *Euler's equation* for an ideal fluid and has to be solved with the equation of conservation of mass, either in the full form (2.3) or the form appropriate to an incompressible fluid, (2.5).

A quantity that is often employed in the discussion of fluid phenomena is the vorticity, $\nabla \times \mathbf{u}$. If the body force $\mathbf{F}$ is conservative, so that it can be derived from a potential (i.e., there exists V such that $\mathbf{F} = -\nabla V$), it is possible to show that
(i) an incompressible ideal fluid started from rest always has zero vorticity
(ii) under some circumstances the vorticity of a viscous fluid decays.

For both reasons the assumption is frequently made that the vorticity of the fluid is zero. Since the equation $\nabla \times \mathbf{u} = \mathbf{0}$ implies the existence of a *velocity potential* ϕ such that $\mathbf{u} = \nabla\phi$,[9] obtaining the solution of problems in ideal fluids is much simplified, and so such an assumption is frequently made. In particular, for an incompressible fluid, it follows that the velocity potential satisfies Laplace's equation

---

[9] Batchelor, 1967, 100.

$$\nabla^2 \phi \equiv \frac{\partial^2 \phi}{\partial x_1^2} + \frac{\partial^2 \phi}{\partial x_2^2} + \frac{\partial^2 \phi}{\partial x_3^2} = 0 . \qquad (2.18)$$

Although it is extremely convenient, it should nonetheless be regarded with a great deal of caution in the modeling of industrial processes of the kind to which thermal modeling is often applied. Frequently, the fluid motion may not be started from rest in the conventional sense, and the time scales are frequently too short for the damping effects of viscosity to have a major effect.

For a fluid of uniform, constant density in steady motion under the influence of body forces derivable from a potential V, the scalar product of Equation (2.17) with $\mathbf{u}$ shows that

$$\mathbf{u}.\nabla\left(\tfrac{1}{2}\rho \mathbf{u}^2 + p + \rho V\right) = 0 .$$

The operator $\mathbf{u}.\nabla$ represents differentiation in the direction of the local velocity field; curves in space that are everywhere tangential to the velocity field are called *stream lines*, and so this equation shows that

$$\tfrac{1}{2}\mathbf{u}^2 + \frac{p}{\rho} + V = C ; \qquad (2.19)$$

C is a constant for any given stream line under those circumstances. This result is *Bernoulli's equation* and can be useful in determining the pressure at different points in the fluid. The constant C is, in general, different for each stream line, although when the motion is also irrotational it is the same for all. It is worth noting that in the special case when the constitutive equation of the fluid takes the form that the pressure is a function of the density only, Bernoulli's equation takes the form

$$\tfrac{1}{2}\mathbf{u}^2 + \int \frac{dp}{\rho} + V \qquad (2.20)$$

is constant on a stream line. This is the case for the adiabatic motion of a gas, for example.

2.1.6 The Navier-Stokes equations for a viscous fluid

The basic assumption of a Newtonian fluid is that the stress tensor has the form

$$p_{ij} = -p\delta_{ij} + d_{ij}$$

where d_{ij}, the deviatoric stress tensor, is an isotropic linear tensor function of the velocity gradient tensor $\dfrac{\partial u_i}{\partial x_j}$, that is to say, it has no directional properties other than those associated with the velocity gradient, and the dependence is linear. This means[10] that the appropriate tensor structure can only be formed from the velocity gradient and the Kronecker delta.[11] The relation has to be constructed in a way that is symmetric in order to satisfy condition (2.16), while p is defined[12] so that $p_{jj} = -3p$. It follows that

$$d_{jj} = 0 \text{ and } d_{ij} = d_{ji}.$$

The most general way to write such an expression with all these properties is

$$d_{ij} = \mu\left(\frac{\partial u_i}{\partial x_j} + \frac{\partial u_j}{\partial x_i} - \frac{2}{3}\delta_{ij}\frac{\partial u_k}{\partial x_k}\right) \equiv 2\mu\left(e_{ij} - \tfrac{1}{3}e_{kk}\delta_{ij}\right)$$

where

$$e_{ij} = \frac{1}{2}\left(\frac{\partial u_i}{\partial x_j} + \frac{\partial u_j}{\partial x_i}\right) \tag{2.21}$$

is the rate of strain tensor for fluid motion. Consequently,

---

[10] Or see Frenkel, 1972.
[11] G.K. Batchelor, 1967, 142-147, and Jeffreys, 1957.
[12] This is not the same definition of pressure as that used in thermodynamics. If that definition is used, a second coefficient of viscosity, the bulk viscosity, would have to be introduced, multiplying $\delta_{ij}e_{kk}$; see (2.21).

$$p_{ij} = -p\delta_{ij} + \mu\left(\frac{\partial u_i}{\partial x_j} + \frac{\partial u_j}{\partial x_i} - \frac{2}{3}\delta_{ij}\frac{\partial u_k}{\partial x_k}\right). \qquad (2.22)$$

Here, μ is the (*dynamical*) *viscosity* of the fluid and Equation (2.22) gives the stress/rate-of-strain relations for a Newtonian fluid. It will be noticed that if the fluid is incompressible, the last term in the bracket vanishes by Equation (2.5). These relations were obtained by Saint-Venant (1843) and Stokes (1845) in essentially the way described here, but had already been derived by Navier (1822) and Poisson (1829) by consideration of the molecular mechanism of internal friction.[13]

It follows that Equation (2.14), when written in vector notation, becomes

$$\rho\frac{\partial \mathbf{u}}{\partial t} + \rho\mathbf{u}.\nabla\mathbf{u} = \rho\mathbf{F} - \nabla p + \mu\left\{\nabla^2\mathbf{u} + \tfrac{1}{3}\nabla(\nabla.\mathbf{u})\right\} \qquad (2.23)$$

for a compressible fluid, or

$$\rho\frac{\partial \mathbf{u}}{\partial t} + \rho\mathbf{u}.\nabla\mathbf{u} = \rho\mathbf{F} - \nabla p + \mu\nabla^2\mathbf{u} \qquad (2.24)$$

in the case of an incompressible fluid. These are the Navier-Stokes equations. They have to be solved in conjunction with the equation of conservation of mass and appropriate boundary conditions. Because of the structure of the equations, the ratio μ/ρ appears very frequently with the result that it is often given the name kinematic viscosity and the symbol ν. The word "kinematic" is often omitted, but it will usually be clear from the context whether μ or ν is intended. If necessary, the term "viscosity," represented here by μ but often by η, can be prefixed by the word "dynamic" or "dynamical" to avoid confusion.

2.1.7 Equations of linear thermoelasticity

Suppose the underlying material considered is an elastic solid that is capable of expansion on being heated. Initially, it is at a uniform temperature T_0 and is at rest. The assumption will be made that under

---

[13] Batchelor, 1967,144-5.

these conditions it has no internal stresses. If that is so, then the stress tensor $\underline{\underline{p}} \equiv 0$. Its elements are then subject to small displacements $\xi(\mathbf{r},t)$ from their original positions as the result of the application of external forces or the existence of a temperature distribution $T(\mathbf{r},t)$. Either of these effects will produce a stress distribution, the first as a result of Hooke's law[14] and the second because of changes in volume as a result of thermal expansion. If the displacements and the temperature difference $T - T_0$ are sufficiently small, the relation between stress and the displacements and temperature field will be linear and additive. Any additional quantities that appear will be either constants or linear isotropic tensors if the underlying material is itself isotropic – that is to say, its properties are independent of orientation. A material for which that would not be the case can be illustrated by any attempt to construct a global model of a reinforced concrete structure. The reinforcing rods would result in the material behaving differently when forces are applied in one direction from the way it would behave if the same forces were applied in another direction. The two effects, distortion and thermal expansion, can be considered separately and then added as a result of the linearity assumption.

Consider first the effects of distortion. Hooke's law states that the extension of a spring is proportional to the force applied. In the case of a continuous body of material, this is equivalent to saying that the relative degree of deformation of adjacent elements of the material depends linearly on the resultant stresses. These relative deformations are measured by the space gradients of the displacement vector ξ, not by ξ itself. To see this, consider the case of a body that is moved rigidly a small amount: no internal stresses are needed to maintain this new position since all elements are still in the same position relative to each other. Thus, there must be a linear isotropic relation between $\dfrac{\partial \xi_i}{\partial x_j}$ and $\underline{\underline{p}}$ that is symmetric, as required by Equation (2.16). The most general such function of p_{ij}, which may only be constructed from p_{ij} and δ_{ij},

---

[14] Named after Robert Hooke, 1635-1703, contemporary and rival of Newton. His Law was first published as an anagram (Hooke, 1676). Its key was published two years later (Hooke, 1678). For an account of the anagram and its meaning see Chapman, 1996.

is a constant scalar multiple of $p_{ij} + v(p_{ij} - p_{kk}\delta_{ij})$ where v is a dimensionless material constant that must not be confused with the kinematic viscosity of a Newtonian fluid. Clearly, it is only the symmetric part of $\dfrac{\partial \xi_i}{\partial x_j}$ that can be expressed in terms of p_{ij}. This is in fact reasonable since the equation obtained by setting the symmetric part equal to zero, i.e., solving

$$e_{ij} \equiv \frac{1}{2}\left(\frac{\partial \xi_i}{\partial x_j} + \frac{\partial \xi_j}{\partial x_i}\right) = 0,$$

has the solution $\xi = c + \omega \times r$, which is a combination of a rigid translation and a rigid rotation, neither of which requires an internal stress distribution in most materials.[15] The tensor e_{ij} is known as the strain tensor and the most general relation between it and p_{ij} is therefore

$$Ee_{ij} = p_{ij} + v(p_{ij} - p_{kk}\delta_{ij}). \tag{2.25}$$

The constant E is Young's modulus ($\mathrm{N\,m^{-2}}$, Pa) for the material, and v is Poisson's ratio. The latter is dimensionless and normally only has values between 0 and $\frac{1}{2}$. The apparently rather strange way in which the right-hand side of (2.25) is defined is for consistency with the standard definition of Young's modulus for the extension of a uniform elastic rod of cross-sectional area A. See Figure 2.3.

If the rod is extended in the direction of the x_1-axis by a force F so that its length changes from h to $h + \delta h$, then the only nonzero component of $\underline{\underline{p}}$ is p_{11}. Its value is F/A, while the component e_{11} of the strain tensor $\underline{\underline{e}}$ is uniform with a value of $\delta h/h$. The standard definition of Young's modulus E requires that

$$\frac{F}{A} = E\frac{\delta h}{h} \quad \text{or} \quad p_{11} = Ee_{11}, \tag{2.26}$$

---

[15] This is an example of the "Principle of Material Frame Indifference."

which is consistent with (2.25). From the same equation, the only other two nonzero components of $\underset{=}{e}$ are e_{22} and e_{33}, and

$$e_{22} = e_{33} = -v e_{11}. \tag{2.27}$$

It is therefore possible to relate Poisson's ratio to the contraction in radius of a circular rod, as will be seen after a consideration of volume changes that occur under the action of a stress distribution.

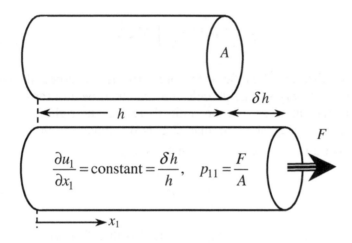

Figure 2.3. Circular rod extended by an applied force.

Consider what happens in Equation (2.25) if i is set equal to j and is summed from 1 to 3. The equation

$$e_{jj} = \frac{\partial \xi_j}{\partial x_j} = \frac{(1-2v)}{E} p_{jj}$$

is obtained, the left-hand side of which is a measure of dilation. Interpretation of $\nabla.\xi$ as a dilation, or change of volume per unit volume, can be seen as follows. Consider a mass of material of initial volume V contained inside the hypothetical surface S with no displacement. Its volume is simply V. Now suppose that its elements are displaced by an amount ξ. So long as the displacements are small, the change in volume of the material can be estimated by finding the new position of the material surface, estimating its thickness at each point, and summing over the whole surface. The new volume is then

$$V + \int_S \xi.\mathbf{n} \, dS$$

since $\xi.\mathbf{n}$ on S is the displacement of a surface element in the normal direction. The ratio of the change in volume to the original volume is therefore

$$\frac{1}{V}\int_S \xi.\mathbf{n} \, dS = \frac{1}{V}\int_V \nabla.\xi \, dV$$

by Gauss's theorem.[16] So far as this argument is concerned, the body of material is arbitrary so that it could, for example, be arbitrarily small in the mathematical sense. In the limit, therefore, the right-hand side of the equation is $\nabla.\xi$, showing that $\nabla.\xi$ is the change in volume of the material per unit volume, showing that, locally,

$$\frac{\delta V}{V} \approx \nabla.\xi = e_{jj} \tag{2.28}$$

with equality in the limit. Equation (2.25) therefore gives the strain tensor in terms of a dilational contribution described in terms of the stress tensor (the first term on the right), and a contribution in which there is no dilation (the second group of terms).

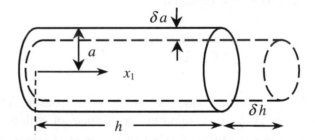

Figure 2.4. Circular rod extended by an applied force.

[16] Kreyszig, 1993, Ch.9.

It is now possible to return to the example shown in Figures 2.3 and 2.4 to see how Poisson's ratio is related to the contraction of a circular rod of radius a. If V is the volume of the rod and δV is its change in volume, then

$$\frac{\delta V}{V} = \frac{\pi(h + \delta h)(a - \delta a)^2 - \pi h a^2}{\pi h a^2}$$

$$\approx \frac{\delta h}{h} - 2\frac{\delta a}{a}. \tag{2.29}$$

Equations (2.26) to (2.28) show that the volume dilation is given by

$$\frac{\delta V}{V} = e_{jj} = (1 - 2v)e_{11} = \frac{(1 - 2v)}{E}p_{11} = (1 - 2v)\frac{F}{AE}$$

where $A = \pi a^2$ is the cross-sectional area of the rod and

$$\frac{\delta h}{h} = \frac{F}{AE}.$$

Equation (2.29) shows that

$$\frac{\delta a}{a} = v\frac{F}{AE},$$

from which the usual definition of Poisson's ratio is obtained.

Equation (2.25) is the stress-strain relation of the classical theory of linear elasticity. Although they can be used with a suitable version of Equation (2.14), they have to be solved for $\underline{\underline{p}}$. Before this is done the effect produced by a temperature distribution must be considered.

A change of temperature alone can introduce a stress distribution, even if there is no displacement at all. It is clear that this is so if one considers what would happen to a body of material that is confined in a rigid container whose shape does not change with temperature. Suppose the material is raised from one uniform temperature to another. Since it is confined within a fixed volume and conditions are uniform, no movement is possible. Because of the tendency to thermal expansion of material when unconfined, however, its elastic compressibility must compensate for this tendency, resulting in an internal stress distribution.

The analysis, however, is more easily performed in relation to the strain tensor than the stress tensor since there is a direct connection between displacement and thermal expansion. Suppose all effects are small enough for the linear approximation to be valid and the material is isotropic. The connection between the strain tensor and the displacement is then most easily considered in a situation where stress is absent and free expansion is allowed. In that case, the strain tensor must be expressible as an isotropic function of the temperature change, so that

$$e_{ij} = \tfrac{1}{3}\alpha_V\left(T - T_0\right)\delta_{ij}. \tag{2.30}$$

If i is set equal to j and this set of equations is summed over j, a process often referred to as contraction, the dilation is obtained, giving

$$e_{kk} = \alpha_V\left(T - T_0\right)$$

so that

$$\alpha_V = \frac{e_{kk}}{\left(T - T_0\right)} = \frac{\delta V}{V}\frac{1}{T - T_0} = \left.\frac{\partial e_{kk}}{\partial T}\right|_{p=0};$$

α_V is therefore the volume coefficient of expansion ($\mathrm{m^3\ m^{-3}\ K^{-1}}$, or just $\mathrm{K^{-1}}$).[17] Its formal definition is

$$\alpha_V = \frac{1}{V}\frac{dV}{dT} = \frac{d\ln V}{dT};$$

the volume coefficient of expansion is three times the linear coefficient $\alpha_L = d\ln\ell/dT$, which is often quoted ($\mathrm{m\ m^{-1}\ K^{-1}}$, or just $\mathrm{K^{-1}}$); ℓ is the length of the specimen. The reason is as follows. Suppose the volume V is a rectangular block with sides ℓ_1, ℓ_2, ℓ_3 so that $V = \ell_1\ell_2\ell_3$, then

---

[17] It will always be assumed here that the units used for the change in volume are the same as the units used for the volume of the initial state.

$$\alpha_V = \frac{d\ln V}{dT} = \sum_{i=1}^{3} \frac{d\ln \ell_i}{dT} = 3\alpha_L.$$

From here on the symbol α will be assumed to refer to the volume coefficient unless otherwise stated.

The linearity assumption means that Equations (2.25) and (2.30) can be combined to give the following stress-strain relations for the linear theory of thermoelasticity,

$$Ee_{ij} = p_{ij} + v\left(p_{ij} - p_{kk}\delta_{ij}\right) + \tfrac{1}{3}E\alpha\delta_{ij}\left(T - T_0\right). \qquad (2.31)$$

This form of the relations is useful for many purposes and is preferred in many engineering contexts. However, it is not the most convenient way of writing them when they need to be used in conjunction with some form of the governing dynamical Equation (2.14). From the form of the equation it is clear that it would be preferable to have the stress tensor given explicitly in terms of the strain tensor. Such an expression can be obtained by the following method. First of all, contract the suffixes i and j to obtain the dilation, from which it follows that

$$p_{kk}\left(1 - 2v\right) = E\left[e_{kk} - \alpha(T - T_0)\right]. \qquad (2.32)$$

From here it can be seen that the dilation is zero in the absence of a temperature change if $v = \tfrac{1}{2}$, which corresponds to the case of an incompressible elastic solid. If (2.32) is substituted into Equation (2.31), it is possible to solve for the stress tensor explicitly, obtaining

$$p_{ij} = \lambda e_{kk}\delta_{ij} + 2\mu\, e_{ij} - \tfrac{1}{3}(3\lambda + 2\mu)\alpha(T - T_0)\delta_{ij} \qquad (2.33)$$

where

$$\lambda_L = \frac{vE}{(1 - 2v)(1 + v)}, \quad \mu_L = \frac{E}{1 + v} \qquad (2.34)$$

or

$$E = \frac{\mu_L(3\lambda_L + 2\mu_L)}{\lambda_L + \mu_L}, \quad v = \frac{\lambda_L}{2(\lambda_L + \mu_L)}; \qquad (2.35)$$

λ_L and μ_L are the Lamé constants. The symbols λ and μ are conventional, but the subscript L will be used here to avoid confusion with the thermal conductivity, λ, and fluid viscosity, μ.

With the stress-strain relations in the form given by (2.33) it is now possible to write down the equation of motion derived from (2.14) for the case of linear thermoelasticity. If it is assumed that the displacements are all small, the nonlinear convective terms must be neglected, and the velocity $\mathbf{u}$ is given by

$$\mathbf{u} = \frac{\partial \xi}{\partial t}$$

while

$$\left(\nabla \cdot \underline{\mathbf{p}}\right)_i = \frac{\partial p_{ji}}{\partial x_j} =$$

$$= \lambda_L \frac{\partial}{\partial x_i}\left(\frac{\partial \xi_k}{\partial x_k}\right) + \mu_L \frac{\partial}{\partial x_j}\left(\frac{\partial \xi_i}{\partial x_j} + \frac{\partial \xi_j}{\partial x_i}\right) - \tfrac{1}{3}(3\lambda_L + 2\mu_L)\alpha\frac{\partial T}{\partial x_i}$$

$$= \left((\lambda_L + \mu_L)\nabla(\nabla\cdot\xi) + \mu_L\nabla^2\xi - \tfrac{1}{3}(3\lambda_L + 2\mu_L)\alpha\nabla T\right)_i$$

so that the equation can be written

$$\rho\frac{\partial^2\xi}{\partial t^2} = \rho\mathbf{F} + (\lambda_L + \mu_L)\nabla(\nabla\cdot\xi) + \mu_L\nabla^2\xi - \tfrac{1}{3}(3\lambda_L + 2\mu_L)\alpha\nabla T. \quad (2.36)$$

This is Navier's equation in a form generalized for thermoelastic problems.

The discussion of the stress-strain relations was presented in terms of small displacements. However, the argument depends only on the stress and strain tensors being small, so it is possible to superimpose a global translation and rotation without changing the value of the strain tensor. These may be time dependent, but if a displacement $\int\{\mathbf{U}(t) + \boldsymbol{\omega}(t)\times\mathbf{r}\}dt$ is superimposed there is no change in the strain tensor. In a great many applications in the theory of material processing, it is more convenient to use a frame of reference for the co-ordinates that is fixed relative to the power source rather than to the workpiece. In that case, the same theory applies so long as the displacements relative to the specified global displacement of the work-piece are small. In a fully time-dependent problem, the connection between velocity at a particular point, and the displacement of the material element that would be at that point if the motion were that of a rigid body, is somewhat complicated. However, if these displacements are small and the velocity of the element to a first approximation is $\mathbf{U}(t)$ with no superimposed rotation, then to the next approximation it is $\mathbf{U}(t) + \left(\dfrac{\partial}{\partial t} + \mathbf{U}.\nabla \right) \xi$. Consequently, the equivalent form to Equation (2.36) is

$$\rho \left\{ \frac{\partial^2 \mathbf{U}}{\partial t^2} + \left(\frac{\partial}{\partial t} + \mathbf{U}.\nabla \right)^2 \xi \right\} =$$

$$\rho \mathbf{F} + (\lambda_L + \mu_L)\nabla(\nabla.\xi) + \mu_L \nabla^2 \xi - \tfrac{1}{3}(3\lambda_L + 2\mu_L)\alpha\nabla T.$$

$$(2.37)$$

In most applications, the first term can be neglected since the acceleration of the workpiece is usually small, but in advanced applications such as laser deposition it might have to be taken into account. It is worth pointing out that in a great many applications the terms on the left can be ignored, although this must always be verified by consideration of orders of magnitude. Thus, for example, in a process described in a coordinate system in which conditions can be considered as steady, the ratio of the inertial term (second on the left) to the elastic terms (second and third groups on the right) is

$$\rho U^2 / E.$$

Typical values for ρ and E for a number of materials can be found in the table in Appendix 1. For example, for the scabbling of concrete with a scanning speed of 10 cm s^{-1} this ratio is of the order of 10^{-10}, and for steel at the same speed it is also of this order of magnitude. Under such circumstances, the inertial term can be neglected entirely.

If the inertial and accelerations terms are neglected, Equation (2.14) takes the rather simpler form

$$\frac{\partial p_{ji}}{\partial x_j} = -\rho F_i \qquad\qquad (2.38)$$

where p_{ij} is given by (2.31), or by (2.33). Differentiation of the definition of e_{ij} shows that

$$\nabla^2 \xi_i = 2\frac{\partial e_{ij}}{\partial x_j} - \frac{\partial e_{jj}}{\partial x_i} \quad \text{and} \quad \nabla^2 e_{ij} = \frac{1}{2}\left(\frac{\partial}{\partial x_j}\nabla^2 \xi_i + \frac{\partial}{\partial x_i}\nabla^2 \xi_j\right).$$

From the definition of the strain tensor

$$e_{ij} \equiv \frac{1}{2}\left(\frac{\partial \xi_i}{\partial x_j} + \frac{\partial \xi_j}{\partial x_i}\right)$$

it is easy to verify the compatibility relations,

$$\frac{\partial^2 e_{ij}}{\partial x_k \partial x_l} + \frac{\partial^2 e_{kl}}{\partial x_i \partial x_j} = \frac{\partial^2 e_{ik}}{\partial x_j \partial x_l} + \frac{\partial^2 e_{jl}}{\partial x_i \partial x_k} \tag{2.39}$$

as both sides are equal to

$$\frac{1}{2}\left(\frac{\partial^3 \xi_i}{\partial x_j \partial x_k \partial x_l} + \frac{\partial^3 \xi_j}{\partial x_i \partial x_k \partial x_l} + \frac{\partial^3 \xi_k}{\partial x_j \partial x_i \partial x_l} + \frac{\partial^3 \xi_l}{\partial x_j \partial x_k \partial x_i}\right).$$

Hence

$$\nabla^2 e_{ij} + \frac{\partial^2 e_{kk}}{\partial x_i \partial x_j} = \frac{\partial^2 e_{ik}}{\partial x_j \partial x_k} + \frac{\partial^2 e_{jk}}{\partial x_k \partial x_i},$$

a result that is obtained by contraction of k and l. Equation (2.31) can be used to eliminate the strain tensor and Equation (2.38) to eliminate $\partial p_{ji}/\partial x_j$. This yields an equation in terms of the stress tensor $\underline{\underline{\mathbf{p}}}$. If that, in turn, is contracted, the contracted form may be used to eliminate $\nabla^2 p_{kk}$, from which it follows that

$$\nabla^2 p_{ij} + \frac{1}{1+v}\frac{\partial^2 p_{kk}}{\partial x_i \partial x_j} = -\frac{v}{1-v}\delta_{ij}\frac{\partial(\rho F_k)}{\partial x_k} - \left\{\frac{\partial(\rho F_i)}{\partial x_j} + \frac{\partial(\rho F_j)}{\partial x_i}\right\}$$

$$-\frac{E\alpha}{3(1-v)}\delta_{ij}\nabla^2 T - \frac{E\alpha}{3(1+v)}\frac{\partial^2 T}{\partial x_i \partial x_j}.$$

$$(2.40)$$

These are the Beltrami-Michel equations and, in general, have to be solved in conjunction with (2.38). They allow problems to be solved directly in terms of the stress tensor, which is often the quantity of more fundamental interest in material processing problems. The alternative is to use the Navier Equation (2.36) for the displacements, and then derive the stress distribution from the stress-strain relations and the definition of the strain tensor.

A particularly simple form is taken in the case of a steady problem in which the components of the stress and strain tensors and the temperature have no dependence on the x_2 coordinates. Then, $p_{12} = p_{32} \equiv 0$ and the body forces are derivable from a potential so that $\rho\mathbf{F} = -\nabla\Omega$ where Ω also has no dependence on x_2. Then Equation (2.38) becomes

$$\frac{\partial}{\partial x_1}(p_{11} - \Omega) + \frac{\partial p_{13}}{\partial x_3} = 0$$

$$\frac{\partial p_{13}}{\partial x_1} + \frac{\partial}{\partial x_3}(p_{33} - \Omega) = 0.$$

From the first two there exists a function χ with the property that

$$p_{11} = \Omega + \frac{\partial^2 \chi}{\partial x_3^2}, \quad p_{13} = -\frac{\partial^2 \chi}{\partial x_1 \partial x_3}, \quad p_{33} = \Omega + \frac{\partial^2 \chi}{\partial x_1^2}; \quad (2.41)$$

χ is the Airy stress function. From the stress-strain relations (2.31)

$$Ee_{11} = p_{11} - vp_{22} - vp_{33} + \tfrac{1}{3}E\alpha\delta_{ij}(T - T_0), \quad Ee_{12} = 0$$
$$Ee_{22} = -vp_{11} + p_{22} - vp_{33} + \tfrac{1}{3}E\alpha\delta_{ij}(T - T_0), \quad Ee_{13} = (1 + v)p_{13}$$
$$Ee_{33} = -vp_{11} - vp_{22} + p_{33} + \tfrac{1}{3}E\alpha\delta_{ij}(T - T_0), \quad Ee_{23} = 0.$$

Two slightly different special cases can be considered. The first is that of *plane strain*, in which the assumption is made that $e_{22} \equiv 0$. The only one of the compatibility relations (2.39) that is not identically zero is

$$\frac{\partial^2 e_{11}}{\partial x_3^2} + \frac{\partial^2 e_{33}}{\partial x_1^2} = 2 \frac{\partial^2 e_{13}}{\partial x_1 \partial x_3}.$$

Combining these equations shows that

$$p_{22} = \nu \nabla^2 \chi + 2\nu \Omega - \tfrac{1}{3} E\alpha(T - T_0), \tag{2.42}$$

$$\nabla^4 \chi + \nabla^2 \left\{ \frac{1 - 2\nu}{1 - \nu} \Omega + \frac{E\alpha}{3(1 - \nu)}(T - T_0) \right\} = 0. \tag{2.43}$$

The case of *plane stress* is very similar, except that a solution is sought in which it is required that $p_{22} \equiv 0$ instead of e_{22}. It then follows that

$$\nabla^4 \chi + \nabla^2 \left\{ (1 - \nu)\Omega + \tfrac{2}{3} E\alpha(T - T_0) \right\} = 0. \tag{2.44}$$

To solve (2.43) or (2.44) it is only necessary to find a particular solution for χ and add to it the general solution for the two-dimensional biharmonic equation

$$\left(\frac{\partial^2}{\partial x_1^2} + \frac{\partial^2}{\partial x_3^2} \right)^2 \chi = 0.$$

The general solution of this is given by Goursart's solution

$$\chi = \text{Re}\{(x_1 - ix_3)f(x_1 + ix_3) + g(x_1 + ix_3)\},$$

which makes it relatively easy to construct analytical solutions of such problems. Goursart's solution[18] is easily verified, and can be shown to be general by rewriting the biharmonic equation in terms of $x_1 + ix_3$ and $x_1 - ix_3$.

---

[18] Muskhelishvili, 1953, Part II Ch. 5.

The cases of plane strain and plane stress are appropriate to rather different circumstances. Quasi two-dimensional models in which a lack of dependence on a coordinate perpendicular to the plane defined by the direction of motion, and the normal to the surface of the workpiece, are usually approximations to three-dimensional models in which the relative configuration is constrained globally. Lateral strain can then only occur with difficulty; in order to maintain the configuration, a lateral stress distribution is needed so that $e_{22} \approx 0$ and $p_{22} \neq 0$. In this case the plane strain model is appropriate. By contrast, a thin plate unconstrained on its surfaces may be assumed to have very little variation in the stresses perpendicular to its plane, but there may be non-zero strains so that the plane stress approximation might be more appropriate.

2.1.8 Plasticity

Normally, the deformation of a material such as a metal satisfies the theory of infinitesimal elasticity for sufficiently small values of the strain. So, if a cylindrical rod is extended longitudinally in the manner shown in Figure 2.2, the elastic behavior for sufficiently small extensions obeys Hooke's law $p = Ee$. In such experiments e is usually defined as $\delta h / h$, where h is the length of the rod and δh is the extension from its original length. Then, p is defined as F/A where F is the force applied and A is the area of cross-section of the rod. The law only holds however if the strain e remains below a critical value, e^*, say. The value of e^*, which depends on the material in question, is of the order of 5×10^{-3} for hardened steels but can be smaller; for annealed metals, for example, it is nearer 10^{-3}. For $e < e^*$ the rod returns to its original length when the stress is removed, and this is also true for $e^* < e < e_y$ where, however, the stress increases more gradually than Hooke's law would allow. The value of the stress corresponding to e_y is known as the yield stress, p_y. For even greater values of e permanent extension occurs and the stress reaches a limiting value.

When the strain reaches a critical value e_f corresponding to the fracture stress p_f, failure occurs and the rod breaks. The value of e_f depends on the metal but can have any value up to about 0.4. Figure 2.5 shows a schematic diagram of such a stress-strain relation for a simple material. It is characteristic that

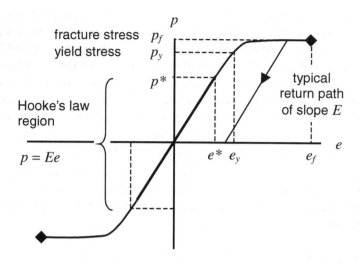

Figure 2.5. Schematic diagram of a one-dimensional stress-strain relation for a simple material.

$$0 \le \frac{\partial p}{\partial e} \le E, \text{ and } \frac{\partial^2 p}{\partial e^2} \le 0 \text{ when } e > 0.$$

If the rod is extended beyond e_y to e_c, for example, where $e_y < e_c < e_f$ and the extending force is then gradually removed, the rod contracts following a stress-strain relation that is nearly linear. It can be approximated by $p = E(e - e_p)$. When the extending force is removed there is a *permanent* or *residual plastic strain* e_p. On repetition of the cycle the new law is followed so long as $e < e_c$. Once greater values of e are reached, either the rod fractures or a new value of e_c is defined with a new value for e_p.

Similar properties hold when the rod is subject to compression. For most metals the complete curve is nearly antisymmetrical.

The stress-strain relation is typical of the particular material and depends on a number of features. One such factor is the chemical composition, but it will also depend on such things as heat treatment and the way in which the object was manufactured. It is therefore not always easy to produce reliable models of industrial processes in which plasticity occurs.

Various simplified models of plasticity have been proposed in the past. One of the simplest is to approximate the kind of stress-strain relation shown in Figure 2.5 by a series of straight line segments as in Figure 2.6.

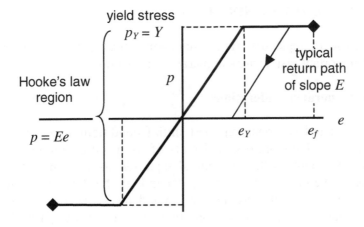

Figure 2.6. Idealized stress-strain relation.

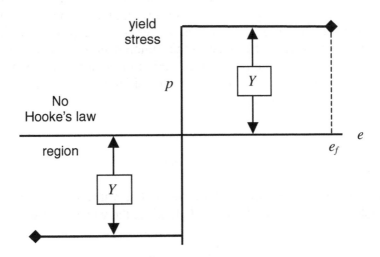

Figure 2.7. Relation for a perfectly plastic solid.

For some purposes it may even be adequate to approximate the relation by the step function shown in Figure 2.7. The generalization of these ideas into three dimensions is not unique, and some additional assumptions are needed. There are two main problems. The first is a suitable criterion for the initiation of yield, and the second is to find an appropriate equation for plastic deformation once yield has occurred.

2.2 BOUNDARY CONDITIONS

This section also uses rather more sophisticated mathematics and the most important results are summarized in Section 2.3.

2.2.1 General considerations

The boundary conditions and initial conditions associated with a particular problem are essential parts of its specification and are every bit as important as the differential equation itself. The number and nature of the conditions normally depends on the type of equation. A full discussion of the problem is beyond the scope of this book, but some mention of the types of conditions associated with particular types of problems is essential.

Conditions to be imposed in problems of the kind studied are of three types. The first is associated with initial value problems. These are problems in which all quantities are in an initial state that is specified. Usually all this requires is the initial values of the dependent variables and perhaps their rate of change in the region occupied by the material that is under study. The exact specification of what these variables are may need some care. For example, an initial value problem for the equation of heat conduction

$$\frac{\partial T}{\partial t} = \kappa \frac{\partial^2 T}{\partial x^2}$$

only requires the specification of the initial value of the dependent variable $T(x,t)$, so that $T(x,0) = T_0(x)$, while the wave equation

$$\frac{\partial^2 u}{\partial t^2} = c^2 \frac{\partial^2 u}{\partial x^2}$$

normally requires that both the initial value of $u(x,t)$ and its t derivative would have to be given so that $u(x,0)=u_0(x)$ and $u_t(x,0)=\dot{u}_0(x)$, where the subscript t indicates the time derivative.

The majority of the problems considered, however, are either specified in a frame of reference in which a steady state is assumed to be possible, with the result that all the unknowns are independent of time, or else only periodic solutions are considered. A function $f(t)$ is periodic if there is a number $T > 0$ with the property that $f(t+T)=f(t)$ for all t. The smallest such number T is called its *period*, and such functions are said to be *periodic*. The simplest examples of periodic functions are the sine and cosine functions since $\sin(t+2\pi)=\sin(t)$ and $\cos(t+2\pi)=\cos(t)$; they have period 2π. Often, more complicated periodic solutions are constructed by adding together combinations of these with other sine and cosine functions, with periods that are integer multiples of 2π, to give solutions in terms of Fourier series.[19]

The second type of condition is a boundary condition at a fixed boundary. An example of such a condition is provided by a solid slab of metal whose faces are maintained at specified temperatures. The most important feature of this kind of condition is that the location of the boundary is known in advance.

The last type of condition is that which occurs on a boundary whose location is actually a part of the solution of the problem. Typically, there will be more conditions at such a boundary than there would be if it were known in advance. A characteristic example is provided by the boundary that separates the molten material in the welding process from the solid metal. Finding this boundary is an essential part of the problem of determining the width of the weld. It is often not at all a simple matter to specify the conditions or to find the boundary. One only has to consider the problems posed by a eutectic alloy. It may not be at all clear how to define such a boundary. In the case of a simple problem of melting, typical conditions might be that the temperature should be equal to the melting temperature of the metal, and that there

---

[19] Kreyszig, 1993, Ch.5.

should be a discontinuity in the heat flow across the boundary given by the latent heat of fusion.

The discussion that follows looks at the main kinds of boundary conditions that have been employed in the past in the modeling of laser processes. It is certainly not complete, and it has to be recognized that, like the choice of a differential equation, all such conditions represent some level of approximation. The skill of the modeler lies in choosing the level of approximation that makes for a problem that can be solved with the available techniques without excessive difficulty, and yet provides the kind of insight, or the kind of accuracy, that was the original purpose of the model. The kinds of approximations used will depend on the objectives.

2.2.2 Thermal boundary conditions

Suppose that the boundary between two adjacent layers of material can be described as a member of the family of surfaces

$$S(\mathbf{r},t)=C.$$

Without any loss of generality, the boundary itself can be taken to be that particular member for which $C=0$. Then suppose $\mathbf{t}$ is a unit vector parallel to the surface. In that case there is no variation in S in the direct of $\mathbf{t}$, so that $(\mathbf{t}.\nabla)S=0$. Rewrite this as $\mathbf{t}.(\nabla S)=0$. It can be seen from this equation that ∇S is perpendicular to any vector $\mathbf{t}$ tangential to the surface. Consequently, the unit normal to the surface can be written as

$$\mathbf{n}=\nabla S/|\nabla S| \qquad\qquad (2.45)$$

at any point at which $|\nabla S|\neq 0$.

The conservation statement in Section 2.1 remains true even if there are discontinuities in the variables concerned. A change of phase is a particular example of such a discontinuity and can be studied from the same point of view as that embodied in the integral statement, Equation (2.1). This time, however, it is not possible to reduce the statement to the form of a partial differential equation. The discontinuity is usually in the density k of the property Q, and sometimes in q. The statement then becomes a relation between the values of quantities on the two sides of the discontinuity. Suppose that the surface on which the discontinuity occurs is given by $S(\mathbf{r},t)=0$. Suppose further that the surface S enclosing the volume V of the integral statement (2.1) is a (temporarily)[20] fixed but small and fairly thin layer of surface area A and thickness h. It contains a part of the surface of discontinuity at both time t and time $t+\delta t$. The faces of area A on both sides of the interface are required to be perpendicular to the normal $\mathbf{n}$. See Figure 2.8.

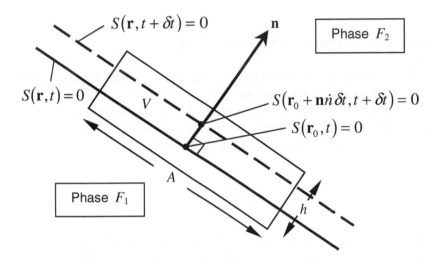

Figure 2.8. The surface enclosing the discontinuity between phases F_1 and F_2.

Now suppose that a point $\mathbf{r}_0$ inside V lies on the interface at time t so that $S(\mathbf{r}_0,t)=0$. If the speed with which the interface moves in the direction of its normal $\mathbf{n}$ is $\dot{n}$, then at a time $t+\delta t$ later the point $\mathbf{r}_0 + \mathbf{n}\dot{n}\delta t$ also lies on the interface, so

$$S(\mathbf{r}_0 + \mathbf{n}\dot{n}\delta t, t + \delta t) - S(\mathbf{r}_0,t) = 0.$$

---

[20] i.e., it is not moving with the interface over the short period of time needed for the ensuing analysis to hold.

Divide by δt and take the limit as $\delta t \to 0$. In the process, make use of the expansion of Taylor's theorem for several variables in the same way as for Equation (2.4). This process shows that

$$\frac{\partial S}{\partial t} + \dot{n}\mathbf{n}.\nabla S = 0,$$

and so[21]

$$\dot{n} = -\frac{\partial S}{\partial t} \Big/ |\nabla S| \tag{2.46}$$

where use has been made of Equation (2.45).

Now consider the consequences of the integral statement (2.1). Suppose the transition is between phases F_1 and F_2 with the normal $\mathbf{n}$ directed from F_1 to F_2. Use subscripts 1 and 2 to distinguish between conditions on the two sides of the interface. Consider each of the terms in (2.1) separately. The principal contribution to

$$\frac{\partial}{\partial t}\int_V \boldsymbol{k}\, dt$$

is the consequence of the change in V of the amount of material in each phase as a result of the change in density. An estimate for this is

$$(\boldsymbol{k}_1 - \boldsymbol{k}_2)A\dot{n}.$$

The remaining contributions to it are of the order of Ah and are therefore an order of magnitude smaller.

In the same way, the major contributions to

$$\int_S \boldsymbol{Q}.\mathbf{n}dS \text{ and } \int_S \boldsymbol{\mathcal{F}}.\mathbf{n}dS$$

come from the integrals over the faces perpendicular to the unit normal $\mathbf{n}$, which are of magnitude A compared to the contributions around the

---

[21] The notation for $\dot{n}$ is convenient, though potentially misleading. It should be remembered that it has the dimensions of a velocity, not a frequency.

sides parallel to it, which are of order of magnitude $h\sqrt{A}$. If the volume is thin in the sense that $h << \sqrt{A}$, then the terms from the integral around the edge can be neglected. If higher order terms are neglected, the largest terms in the estimates for these two integrals therefore become

$$-\boldsymbol{Q}_1.\mathbf{n}A + \boldsymbol{Q}_2.\mathbf{n}A \text{ and } -\boldsymbol{\mathcal{T}}_1.\mathbf{n}A + \boldsymbol{\mathcal{T}}_2.\mathbf{n}A.$$

Lastly, the contribution from q is dominated by any surface effects such as surface tension or latent heat. Suppose this corresponds to a rate of generation Δq units of Q per unit area when the interface is crossed in the same direction as $\mathbf{n}$. Then

$$\int_V q\, dV$$

is approximated by $\Delta q\, A$. Replace the integrals in (2.1) by these approximations, divide by A, and take the limit as $A \rightarrow 0$ in such a way that the maximum dimension of A also tends to zero and that the limit is taken in such a way that h is always small compared to $\sqrt{A}$. The result is the interface condition

$$(k_1 - k_2)\dot{n} - \boldsymbol{Q}_1.\mathbf{n} + \boldsymbol{Q}_2.\mathbf{n} = -\boldsymbol{\mathcal{T}}_1.\mathbf{n} + \boldsymbol{\mathcal{T}}_2.\mathbf{n} + \Delta q$$

or, more concisely,

$$\left[-k\dot{n} + \boldsymbol{Q}.\mathbf{n} - \boldsymbol{\mathcal{T}}.\mathbf{n}\right]_1^2 = \Delta q. \tag{2.47}$$

Equations (2.45) and (2.46) show that an equivalent way of writing this is

$$\left[k\frac{\partial S}{\partial t} + (\boldsymbol{Q} - \boldsymbol{\mathcal{T}}).\nabla S\right]_1^2 = |\nabla S|\Delta q. \tag{2.48}$$

So, for example, in the conservation of mass condition, k is ρ, $\boldsymbol{Q}$ is $\rho\mathbf{u}$, $\boldsymbol{\mathcal{T}}$ is $\mathbf{0}$, and Δq is 0. The interface condition for conservation of mass is therefore

$$\left[\rho(\mathbf{u.n} - \dot{n})\right]_1^2 = 0 \text{ or } \left[\rho \frac{DS}{Dt}\right]_1^2 = 0.$$

See the discussion following Equation (2.3) for the definition of the inertial derivative, D/Dt. The square brackets indicate the change in value of the quantity enclosed between the two sides of the interface, so, for example,

$$[X]_1^2 = X(\text{on side 2 of the interface}) - X(\text{on side 1 of the interface}).$$

Notice that the mass flow rate per unit area $\dot{m}$ across the boundary can be written in any of the forms

$$\dot{m} = \left[\rho(\mathbf{u.n} - \dot{n})\right]^1 = \left[\rho(\mathbf{u.n} - \dot{n})\right]^2 = \left[\frac{\rho}{|\nabla S|}\frac{DS}{Dt}\right]^1 = \left[\frac{\rho}{|\nabla S|}\frac{DS}{Dt}\right]^2 \qquad (2.49)$$

In the same way, consider the case when Q is thermal energy. Then

$$k \text{ is } \rho\int_{T_0}^T c_p dT, \ \mathbf{Q} \text{ is } \rho\mathbf{u}\int_{T_0}^T c_p dT - \lambda\nabla T, \ \mathcal{J} \text{ is } 0 \text{ and } \Delta q \text{ is } -\dot{m}L_1^2$$

where L_1^2 is the latent heat of transition from phase 1 to phase 2, defined to be positive if an input of energy is required as the material moves from side 1 to side 2 of the interface, i.e., in the direction of the normal $\mathbf{n}$. The temperature integrations are from some suitable fixed reference temperature. They are defined in such a way that any discontinuities of the integrand are finite; the contribution from latent heat therefore appears only in the term in L. The integrals themselves are continuous across the interface with the same value on each side. Condition (2.48) shows that

$$\left[\rho(\mathbf{u.n} - \dot{n})\int_{T_0}^T c_p dT - \lambda\mathbf{n}.\nabla T\right]_1^2 = -\dot{m}L_1^2.$$

Since the temperature integral has the same value on each side of the boundary, the mass conservation condition (2.49) shows that the change in value of the first term is zero. Hence, this condition, often known as the *Stefan condition*, can be written as

$$[\lambda \mathbf{n}.\nabla T]_1^2 = \dot{m}L_1^2 \quad \text{or} \quad [\lambda \nabla S.\nabla T]_1^2 = \dot{m}|\nabla S|L_1^2 \tag{2.50}$$

where the mass flow rate $\dot{m}$ can be given by any of the forms shown in (2.49). So, for example, at a melting boundary

$$[\lambda \mathbf{n}.\nabla T]_S^L = \dot{m}L_M \tag{2.51}$$

where $L_M = L_S^L$ is the latent heat of melting.

There is a difficulty associated with the condition at the boundary between the solid and the liquid phase. The boundary may not be clearly defined; since the solid and liquid states are different phases of the same material, it is possible for there to be a region in which solid and liquid forms exist together in thermodynamic equilibrium. The problem is particularly acute in the case of an alloy (as will nearly always be the case in practice) if it is allowed to cool slowly. Some form of separation will occur on freezing, and this can give rise to an intermediate "mushy" zone in which liquid and solid material can coexist in what can be a region of substantial size. The dendritic freezing of a eutectic mixture is a case in point.[22]

In many of the problems considered here, the change of phase is brought about by an input of energy from a laser beam. This is usually the case when the boundary is one at which boiling takes place. If the absorbed normal incident intensity is I, then I needs to be added to the right-hand side of (2.47) since I is, in effect, an addition to the definition of Δq. Consequently, the appropriate form of (2.47) when boiling takes place is

$$[\lambda \mathbf{n}.\nabla T]_L^V = -I + \dot{m}L_B \tag{2.52}$$

where $L_B = L_L^V$ is the latent heat of boiling.

---

[22] Peel, 1975, 5-18. Atthey, 1974, gives a method for tackling such problems in the absence of relative motion between the two states.

2.2.3 Dynamical boundary conditions

When conditions at the boundary between two fluid phases are considered the basic quantity, Q is the i-component of the momentum density vector, ρu_i, in a rectangular Cartesian coordinate system. Then k is ρu_i, Q is $\rho u_i \mathbf{u}$, and $\mathcal{T}$ is $-p\underline{\underline{I}}_i + \underline{\underline{d}}_i$ where $\underline{\underline{I}}$ is the identity diadic and $\underline{\underline{d}}$ is the deviatoric stress tensor.[23] Thus the left-hand side of (2.47) is

$$\left[\rho\mathbf{u}(\mathbf{u.n} - \dot{n}) + p\mathbf{n} - \underline{\underline{d}}.\mathbf{n}\right]_1^2 . \tag{2.53}$$

To find the right-hand side it is necessary to consider further the form that Δq takes. In this instance the generating quantity is surface tension. Suppose the surface tension is γ and consider the force on an element S in the surface whose perimeter is C. See Figure 2.9.

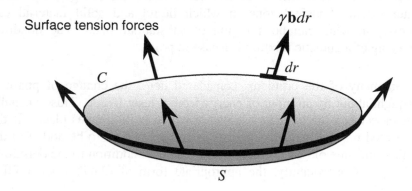

Figure 2.9. The surface tension forces on a portion S of the interface.

The total force on S as a result of the action of surface tension is

$$\int_C \gamma \mathbf{b}\, dr$$

---

[23] See discussion preceding Equation (2.21).

where dr is the length of an element of C and $\mathbf{b}$ is a unit vector perpendicular to C in the tangent plane to S at the point on C under consideration. However, it is possible to relate $\mathbf{b}$ to the vector direction $\mathbf{dr}$ of dr and the local normal, $\mathbf{n}$. The relation is $\mathbf{b}dr = \mathbf{dr} \times \mathbf{n}$ when the usual convention is employed that the direction of description of C is in a right-handed sense relative to the normal $\mathbf{n}$ of S. The direction must, of course, be consistently defined over the whole of S; arbitrary exchanges of direction from one part of S to another are not allowed. It will be assumed that S can be described as a member of a family of surfaces in the manner described above. See Figure 2.10.

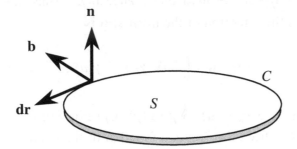

Figure 2.10. The relation between the normal $\mathbf{n}$, the element $\mathbf{dr}$ of C, and the vector $\mathbf{b}$ in the direction of the surface tension force.

Consequently, the net surface tension force $\mathbf{F}$ on S is

$$\int_C \gamma \mathbf{dr} \times \mathbf{n} = -\int_C \gamma \frac{\nabla S \times \mathbf{dr}}{|\nabla S|}. \qquad (2.54)$$

It is possible to transform this integral into a surface integral by means of Stokes' theorem.[24] This integral theorem states that for a sufficiently smooth surface S bounded by a sufficiently smooth curve C, and for a vector field $\mathbf{u}(\mathbf{r})$, which is differentiable on S,

$$\int_C \mathbf{u}.\mathbf{dr} = \int_S \nabla \times \mathbf{u}.\mathbf{n} dS.$$

To see how the theorem can be applied, take the scalar product of (2.54) with a constant vector $\mathbf{a}$ and rearrange the integrand to show that

---

[24] Kreyszig, 1993, Ch.9.

$$\mathbf{a}.\int_C \gamma \, d\mathbf{r} \times \mathbf{n} = -\int_C \left(\gamma \mathbf{a} \times \frac{\nabla S}{|\nabla S|} \right).d\mathbf{r} \ .$$

Apply Stokes' theorem and expand the resulting double cross product using one of the standard vector identities[25] to show that

$$\mathbf{a}.\int_C \gamma \, d\mathbf{r} \times \mathbf{n} = \int_S \left\{ (\mathbf{a}.\mathbf{n})\nabla.(\gamma \mathbf{n}) - \mathbf{n}^2 \mathbf{a}.\nabla \gamma - \tfrac{1}{2}\gamma \nabla \mathbf{n}^2 \right\} dS$$

$$= \mathbf{a}.\int_S \left\{ \mathbf{n}(\nabla.\gamma \mathbf{n}) - \nabla \gamma \right\} dS$$

where $\mathbf{n}$ and $\nabla S / |\nabla S|$ are used interchangeably. Thus, the surface tension force in the direction of the normal $\mathbf{n}$ is

$$\int_C \gamma \, d\mathbf{r} \times \mathbf{n} = \int_S \left\{ \mathbf{n}(\nabla.\gamma \mathbf{n}) - \nabla \gamma \right\} dS \ ,$$

equivalent to a force $\mathbf{n}(\nabla.\gamma \mathbf{n}) - \nabla \gamma = \mathbf{n}(\gamma \nabla.\mathbf{n}) + \mathbf{n}(\mathbf{n}.\nabla \gamma) - \nabla \gamma$ per unit area. It will be noticed that if γ is a function of the temperature only, this can be written

$$\mathbf{n}(\gamma \nabla.\mathbf{n}) + \frac{d\gamma}{dT} \left\{ \mathbf{n}(\mathbf{n}.\nabla T) - \nabla T \right\}. \tag{2.55}$$

Notice that the second group of terms is tangential to the interface and the first is normal to it.

Combine this result with (2.53) and substitute them into (2.47) to show that the interface condition at a fluid boundary is

$$\left[\rho \mathbf{u}(\mathbf{u}.\mathbf{n} - \dot{n}) + p\mathbf{n} - \underline{\underline{d}}.\mathbf{n} \right]_1^2 = \mathbf{n}(\gamma \nabla.\mathbf{n}) + \left\{ \mathbf{n}(\mathbf{n}.\nabla \gamma) - \nabla \gamma \right\}. \tag{2.56}$$

This equation has a component normal to the surface and two components tangential to it. The most usual application is at a liquid/vapor boundary, in which case 2 is replaced by V, and 1 by L.

---

[25] Kreyszig, 1993, Ch.8.

2.2.4 Other conditions

There are some further conditions that are not of conservation type.

Thus, the usual condition at an interface between two different phases is that the temperature should be continuous. So, for example,

$$[T]_S^L = 0 \text{ or } [T]_L^V = 0. \tag{2.57}$$

The same is true at a boundary between different solid phases, as in steel, for example. Some caution is needed, however, in the case of a strongly evaporating boundary where the presence of a Knudsen[26] layer can affect the result.

In the case of a viscous liquid, there will be a no-slip condition on the tangential component of velocity. This will apply whether the interface is between a solid and a fluid region, or between two fluid phases. Thus

$$[\mathbf{u.t}]_1^2 = 0 \tag{2.58}$$

for all vectors $\mathbf{t}$ tangential to the surface,[27] so that $\mathbf{t.n} = 0$.

2.2.5 Comments on the fluid boundary conditions

There are at least two particular problems when it comes to constructing simple analytical models of processes such as welding when a deliberate attempt is made to include a reasonably realistic allowance for fluid motion. One is the difficulty raised by the fact that the location of the boundary between different phases of the metal is itself a part of the solution of the problem; the other is the fact that liquid phases satisfy the equations of fluid dynamics. These, in all but their simplest forms, are nonlinear and are capable of very complicated flows that are not easily described in analytical terms, and are indeed hard to obtain by computational methods. The flow in the vapor phase is often at or near the turbulent regime, and the motion in the weld pool is known from observation to be complicated.[28]

---

[26] See, e.g., Finke and Simon, 1990.

[27] Batchelor, 1967, p.149.

[28] Arata et al., 1976; Arata and Miyamoto, 1978; Matsunawa, 2000.

It can be argued that a linearized version of the Navier-Stokes Equation (2.23), which includes the effects of viscosity, can sometimes be usefully employed. An alternative approach might be to use an adaptation of classical boundary layer theory.[29]   A very fruitful approach to the solution of fluid dynamical problems in the past has been to use the Euler Equation (2.17) in which viscous effects are entirely ignored.   This approximation is generally used in fluid dynamics in connection with the assumption that the flow is irrotational and incompressible, so that there exists a velocity potential ϕ satisfying Laplace's Equation (2.18). A problem with this approximation is that in welding problems, for example, the fluid in question is metal that has only just become molten. The time scale of its subsequent motion may be so short that there is no reason to suppose that viscous forces can damp out any vorticity present. At first sight it might seem plausible that, since the material melts from a state in which the motion is parallel and uniform and thus has no vorticity, then the molten flow will also have no vorticity. That, however, is not necessarily the case. Because most of the material in the molten zone actually flows around the keyhole, there has to be a transverse flow that must be generated by a pressure gradient in the molten region. A simple example will serve to show the appropriate boundary conditions, and it will be seen that the liquid does not clearly satisfy an irrotational condition on melting.

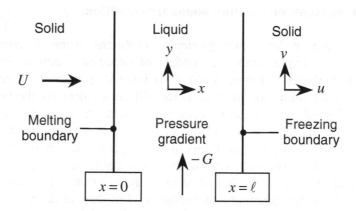

Figure 2.11. Schematic diagram of a simple model of material moving from a solid phase though a liquid phase in which there is a transverse pressure gradient, and returning to the solid phase.

---

[29] Rosenhead, 1963.

Consider a simple two-dimensional model of viscous flow of an incompressible liquid across a channel in which there is a transverse pressure gradient, as shown in Figure 2.11.

The equation satisfied is the steady-state Navier-Stokes Equation (2.24). The x-component of velocity u is equal to U everywhere, and the transverse velocity v satisfies

$$\rho U \frac{dv}{dx} = G + \mu \frac{d^2 v}{dx^2}$$

where G is the transverse pressure gradient. Suppose that the material melts at $x = 0$ and freezes again at $x = \ell$. Boundary conditions at $x = 0$ and $x = \ell$ are $v(0) = v(\ell) = 0$. This is a simple second-order ordinary differential equation whose general solution is $Gx/\rho U$ plus a constant and an arbitrary multiple of $\exp(\rho U x/\mu)$. The only combination that satisfies both boundary conditions is

$$v = \frac{G\ell}{\rho U} \left\{ \frac{x}{\ell} - \frac{\sinh(\rho U x/2\mu)}{\sinh(\rho U \ell/2\mu)} \exp\left[-\frac{(\ell - x)\rho U}{2\mu} \right] \right\}.$$

Consequently, in the limit as $\mu \to 0$,

$$v \to \begin{cases} \dfrac{Gx}{\rho U} & \text{in} \quad 0 \le x < \ell \\ 0 & \text{at} \quad x = \ell. \end{cases}$$

Thus, the natural conditions to apply in such problems are that

  (i) the normal component of mass flux and the tangential component of velocity are continuous at a melting boundary,
  (ii) the normal component of the mass flux is continuous at a freezing boundary.

                                                                    (2.59)

These conditions may lead to irrotational flow in special circumstances, but in general will not. If the irrotational approximation is nonetheless used as a worthwhile simplification of an otherwise intractable problem, it is therefore advisable in principle to estimate the error introduced to check that it is small enough not to invalidate the main conclusions of the particular investigation.

2.2.6 Elastic boundary conditions

Thermoelastic problems in material science can usually be divided into two parts, one with a response time scale of the thermal problem, and one with a purely elastic character and a much more rapid response. Only problems in which conditions are quasi-steady will be considered, so that a representation is used in which there is no explicit dependence on time. Typical boundary conditions are then the same as those for steady, purely elastic problems.

One type of problem is that in which the surface tractions are specified on part or all of the boundary. This constitutes a condition on the stress tensor, so that typically

$$p_{ij} n_j = \tau_i \qquad\qquad (2.60)$$

on the boundary. The normal to the surface is $\mathbf{n}$, and the specified surface traction is τ. On a free boundary, for instance, $\underline{\underline{p}}.\mathbf{n} = \mathbf{0}$.

Sometimes, however, the displacement on the surface may be given so that the condition on those parts of the surface is

$$\xi = \xi_0 \qquad\qquad (2.61)$$

where ξ_0 is the specified displacement. So, for example, a portion of the boundary that is rigidly fixed will satisfy the condition $\xi = \mathbf{0}$.

In some problems, however, a mixed condition may be more appropriate. In laser welding, for example, the workpiece may be clamped. This is necessary for various reasons. One might be to attach it to the work table so that it can be accurately moved relative to the laser; another might be to attempt to minimize deformations. The forces involved can be very substantial so that, although it may be possible to prevent deformations far from the treatment area normal to the work-piece, the clamps may not be able to prevent small but significant lateral deformation. The whole effect of clamping a workpiece does not appear to be well understood, but it may be that a more appropriate set of conditions under some circumstances, rather than (2.60) or (2.61), might be

$$\xi.\mathbf{n} = 0, \; \mathbf{t}.\underline{\underline{\mathbf{p}}}.\mathbf{n} = 0 \; \text{ for all } \mathbf{t}.\mathbf{n} = 0. \tag{2.62}$$

This last condition represents no deformation in the direction held by the clamps, but no effective constraint on lateral motion.

2.3 SUMMARY OF EQUATIONS AND CONDITIONS

2.3.1 General

The purpose here is to list those equations that are most used in the rest of this book. They are not the most general versions of the equations in question. Unless stated otherwise, the forms given are time-independent and the material parameters, such as thermal conductivity, for example, are taken to be constants. The meanings of the symbols employed in each context are given. There really are not enough letters in the Greek and Roman alphabets, however. It is inevitable that some symbols will be used to mean several different things, and different authors may use symbols in different ways. For example, the Greek letter v is commonly used for the kinematic viscosity of a liquid and for the Poisson's ratio of an elastic solid, and in some fonts is almost indistinguishable from italic v – as here. A further complication is provided by the fact that more than one symbol may be used for the same quantity; σ is also often used for Poisson's ratio, for instance. It should be clear from the context what is intended, but the reader needs to be aware of the problems of notation, and to be clear what conventions are being used by the author of the work being read.

Under each heading, standard forms for the equations and boundary conditions are given that are suitable for the kinds of problems considered. Also listed are the symbols used, their SI base units, and their names, together with related and derived symbols. After their names, symbols that are in common use as alternatives are given in square brackets.

The following are common to all the topics considered.

t	s	is time;
(u,v,w)	m s^{-1}	are the coordinates of velocity $[(u_1,u_2,u_3)]$;
(x,y,z)	m	are Cartesian coordinates $[(x_1,x_2,x_3)]$;
(ξ,η,ς)	m	are displacements $[(\xi_1,\xi_2,\xi_3), (u_1,u_2,u_3)]$.

2.3.2 Thermal equations and conditions

From Equation (2.10), the steady-state equation of heat conduction for a workpiece moving with a constant velocity U in the direction of the x-axis when all the material parameters are considered to be constant is

$$\rho\,c_p U \frac{\partial T}{\partial x} = \lambda \left(\frac{\partial^2 T}{\partial x^2} + \frac{\partial^2 T}{\partial y^2} + \frac{\partial^2 T}{\partial z^2} \right) + q \qquad (2.63)$$

and if time dependence and a nonuniform velocity field is included,

$$\rho\,c_p \left(\frac{\partial T}{\partial t} + u\frac{\partial T}{\partial x} + v\frac{\partial T}{\partial y} w\frac{\partial T}{\partial z} \right) = \lambda \left(\frac{\partial^2 T}{\partial x^2} + \frac{\partial^2 T}{\partial y^2} + \frac{\partial^2 T}{\partial z^2} \right) + q\,. \qquad (2.64)$$

At an isothermal boundary, between two different media 1 and 2 (or different phases of the same medium)

$$[T]_1^2 = 0\,. \qquad (2.65)$$

If where L_M is the latent heat of melting, the Stefan condition is

$$[\lambda \mathbf{n}.\nabla T]_S^L = \dot{m}L_M \qquad (2.66)$$

at a melting boundary; $\dot{m}$ is the mass flow rate per unit area normal to the boundary. At a boundary at which boiling takes place,

$$[\lambda \mathbf{n}.\nabla T]_L^V = \dot{m}L_B - I \tag{2.67}$$

where L_B is the latent heat of boiling and I is an incident absorbed intensity, supplied by a laser, for example.

c_p	J kg^{-1} K^{-1}	is the specific heat of the material;
λ	W m^{-1} K^{-1}	is its thermal conductivity [k];
q	W m^{-3}	is the rate of energy absorption;
T	K	is its temperature;
U	m s^{-1}	is a constant velocity of translation in the x-direction [V];
$\kappa = \dfrac{\lambda}{\rho c_p}$	m s^{-2}	is the thermal diffusivity [a];
ρ	kg m^{-3}	is the density of the material;

If

ℓ	m	is a characteristic length scale

then

$\text{Pe} = \dfrac{U\ell}{\kappa}$	–	is the Péclet number.

2.3.3 Fluid motion

From (2.3), the equation of conservation of mass in Cartesian co-ordinates is

$$\frac{D\rho}{Dt} + \rho \nabla.\mathbf{u} = 0 \tag{2.68}$$

where

$$\frac{D}{Dt} \equiv \frac{\partial}{\partial t} + u\frac{\partial}{\partial x} + v\frac{\partial}{\partial y} + w\frac{\partial}{\partial z} \quad \text{and} \quad \nabla.\mathbf{u} \equiv \frac{\partial u}{\partial x} + \frac{\partial v}{\partial y} + \frac{\partial w}{\partial z}.$$

From (2.23), the Navier-Stokes equations for a compressible fluid are

$$\rho \frac{Du}{Dt} = -\frac{\partial p}{\partial x} + \rho F_1 + \mu \left(\nabla^2 u + \frac{1}{3} \frac{\partial}{\partial x} \nabla . \mathbf{u} \right)$$

$$\rho \frac{Dv}{Dt} = -\frac{\partial p}{\partial y} + \rho F_2 + \mu \left(\nabla^2 v + \frac{1}{3} \frac{\partial}{\partial y} \nabla . \mathbf{u} \right) \qquad (2.69)$$

$$\rho \frac{Dw}{Dt} = -\frac{\partial p}{\partial z} + \rho F_3 + \mu \left(\nabla^2 w + \frac{1}{3} \frac{\partial}{\partial z} \nabla . \mathbf{u} \right)$$

where

$$\nabla^2 \equiv \frac{\partial^2}{\partial x^2} + \frac{\partial^2}{\partial y^2} + \frac{\partial^2}{\partial z^2} .$$

The equations for an incompressible fluid are obtained by replacing (2.68) by

$$\nabla . \mathbf{u} = 0 \qquad (2.70)$$

and substituting it in (2.69).

The Euler equations for an ideal fluid (inviscid flow) are obtained by setting $\mu = 0$ in (2.69).

In steady irrotational flow

$$\mathbf{u} = \left(\frac{\partial \phi}{\partial x}, \frac{\partial \phi}{\partial y}, \frac{\partial \phi}{\partial z} \right) \text{ where } \nabla^2 \phi = 0 . \qquad (2.71)$$

- In steady incompressible viscous flow:
  $\mathbf{u} = \mathbf{0}$ at a fixed solid boundary or $\mathbf{u}$ is given at a moving solid boundary;
  $\rho \mathbf{u} . \mathbf{n}$ and $\mathbf{u} . \mathbf{t}$ are continuous when fluid crosses an internal interface (as at a change of phase, for example);
  $- p + 2\mu \mathbf{n} . \{(\mathbf{n} . \nabla) \mathbf{u}\}$ and $\mu [\mathbf{t} . \{(\mathbf{n} . \nabla) \mathbf{u}\} + \mathbf{n} . \{(\mathbf{t} . \nabla) \mathbf{u}\}]$ are continuous for an incompressible liquid at a boundary between two fluids (which may be different phases of the same fluid) ignoring surface tension; see Equation (2.56) for the form of the condition when surface tension is included, and note the effect of the continuity

of **u.t** on the tangential component of the inertial contribution;

- in steady inviscid flow, **u.n** is specified at a solid boundary and is zero if the boundary is at rest or in motion parallel to itself;
- at a melting boundary in steady flow where material melts into an inviscid flow, ρ**u.n** and **u.t** are continuous, but at a freezing boundary only ρ**u.n** is continuous.

In this list of conditions, **n** is the normal to the boundary and **t** is any tangent to the boundary.

(F_1, F_2, F_3)	N kg^{-1}	is the body force per unit mass;
p	N m^{-2}; Pa	is the pressure;
$\Pr = \dfrac{v}{\kappa}$	–	is the Prandtl number;
μ	kg m^{-1} s^{-1}	is the viscosity $[\eta]$;
v	m^2 s^{-1}	is the kinematic viscosity;
ρ	kg m^{-3}	is the density;
ϕ	m^2 s^{-1}	is the velocity potential;

If

ℓ	m	is a characteristic length scale

then

$\mathrm{Re} = \dfrac{U\ell}{v}$	–	is the Reynolds number.

2.3.4 Thermoelasticity

The steady-state equations of thermoelastic flow are

$$\rho\left(\frac{\partial}{\partial t}\mathbf{U}.\nabla\right)^2 \xi = \rho\mathbf{F} + (\lambda_L + \mu_L)\nabla(\nabla.\xi) + \mu_L \nabla^2\xi - \tfrac{1}{3}(3\lambda_L + 2\mu_L)\alpha\nabla T \tag{2.72}$$

(but the first term can usually be neglected) when the material is in motion with a constant velocity **U** independent of position in space.

$$\lambda_L = \frac{vE}{(1-2v)(1+v)}, \quad \mu_L = \frac{E}{1+v}, \tag{2.73}$$

$$E = \frac{\mu_L(3\lambda_L + 2\mu_L)}{\lambda_L + \mu_L}, \quad v = \frac{\lambda_L}{2(\lambda_L + \mu_L)}. \tag{2.74}$$

Characteristic boundary conditions are

(i) $p_{ij}n_j = \tau_i$ (2.75)

on a boundary with specified traction τ.

When the displacements are given

(ii) $\xi = \xi_0$ (2.76)

where ξ_0 is the specified displacement.

E	Pa	Young's modulus;
$\mathbf{F} = (F_1, F_2, F_3)$	N kg^{-1}	is the body force per unit mass;
α	K^{-1}	is the volume coefficient of expansion;
λ_L, μ_L	Pa	are the Lamé constants [λ, μ];
v	–	Poisson's ratio [σ];
ρ	kg m^{-3}	is the density;
ξ	m	is the displacement from the position the element would have occupied in the absence of thermoelastic distortion [**u**].

CHAPTER 3

THE TEMPERATURE IN BLOCKS AND PLATES

3.1 THE TEMPERATURE DISTRIBUTION

In many forms of laser technology the coherent light from the laser forms a spot that can be a concentrated source of heat or, in laser surface treatment, for example, a rather more diffuse region of heating. Some of the techniques used in the simpler mathematical models that first found use in welding problems in fact prove to be of considerable value in more general contexts. In particular, the point and line source solutions associated in the context of welding with the name of Rosenthal,[1] but also very well described from a rather different point of view by Carslaw and Jaeger,[2] have proved to be extremely useful. They will be derived first before considering specific applications, after which they will be used to obtain simple descriptions of the temperature in a workpiece. More elaborate models can be developed later.

From these elementary solutions it is possible to build up more complex solutions to describe different incident intensity distributions at the surface of the workpiece. They can be extended to cover time-dependent situations. The line solution can be applied to a plate of finite thickness as well as an infinite or semi-infinite workpiece. The point solution in its basic form only applies to an infinite or semi-infinite workpiece, but it can be extended very simply to cover other cases, such as a plate of finite thickness, for example.

In various ways it is therefore possible to solve more complicated problems in terms of these simple analytical solutions, an approach that can lead to better understanding before resorting to more complicated computational methods.

Whatever approach is used for the power of the incident radiation

---

[1] Rosenthal, 1941 and 1946.
[2] Carslaw and Jaeger, 1959.

and the energy transfer mechanisms, the result is a prediction for the temperature distribution in the workpiece. It can then be used as the basis for further calculation to obtain such quantities as

- the thermal history of individual points in the workpiece;
- metallurgical properties deduced from the thermal histories;
- more accurate solutions for the temperature distribution, using the point and line source solution as the first stage of an iterative scheme, which might be analytical or, more usually, numerical;
- the consequences of varying the absorption models in order to test their validity and reliability;
- the distribution of thermal stress in the workpiece;
- the deformation of the workpiece resulting from thermal stress.

3.2 THE POINT SOURCE SOLUTION

3.2.1 Special solutions

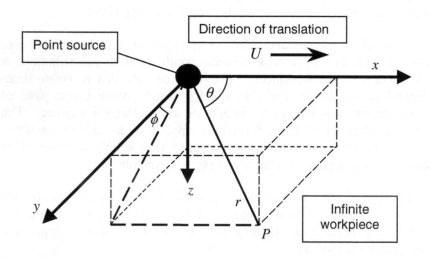

Figure 3.1. The relative geometry and the coordinate system employed in the description of the point source solution. P is the point given by $x = r \cos \theta$, $y = r \sin \theta \cos \phi$, $z = r \sin \theta \sin \phi$ (the point source solution is axisymmetric, so ϕ does not appear explicitly in it).

If it is assumed that the workpiece is in steady motion parallel to the x-axis with velocity U and that all the material parameters are constants, Equation (2.63) shows that the equation of heat conduction for the temperature T takes the relatively simple form

$$U\frac{\partial T}{\partial x} = \kappa\left(\frac{\partial^2 T}{\partial x^2} + \frac{\partial^2 T}{\partial y^2} + \frac{\partial^2 T}{\partial z^2}\right). \tag{3.1}$$

The parameter κ is the thermal diffusivity. See Figure 3.1 for the relative geometry and the coordinate system.

The point and line source models are two special solutions of Equation (3.1) that can be obtained as follows. If the transformation

$$T = T_0 + S\exp\left(\frac{Ux}{2\kappa}\right) \tag{3.2}$$

is employed, where T_0 is the ambient temperature, Equation (3.1) reduces to

$$\frac{\partial^2 S}{\partial x^2} + \frac{\partial^2 S}{\partial x^2} + \frac{\partial^2 S}{\partial x^2} = \frac{U^2}{4\kappa^2}S. \tag{3.3}$$

Now look for a solution for S that depends only on the radial distance from the origin, $r = \sqrt{x^2 + y^2 + z^2}$. In that case, the chain rule shows that

$$\frac{\partial S}{\partial x} = \frac{\partial r}{\partial x}\frac{dS}{dr} = \frac{x}{r}\frac{dS}{dr}.$$

The process can be repeated to show that

$$\frac{\partial^2 S}{\partial x^2} = \frac{1}{r}\frac{dS}{dr} + \frac{x^2}{r}\frac{d}{dr}\left(\frac{1}{r}\frac{dS}{dr}\right) = \frac{1}{r}\frac{dS}{dr} + \frac{x^2}{r^2}\frac{d^2 S}{dr^2} - \frac{x^2}{r^3}\frac{dS}{dr}.$$

The expressions for the second derivatives of S with respect to y and z are identical except that x is replaced by y for $\frac{\partial^2 S}{\partial y^2}$ and by z for $\frac{\partial^2 S}{\partial z^2}$.

Adding them together shows that

$$\frac{d^2 S}{dr^2} + \frac{2}{r}\frac{dS}{dr} = \frac{U^2}{4\kappa^2} S.$$

It is now a straightforward matter to verify that this is the same equation as

$$\frac{d^2}{dr^2}(rS) = \frac{U^2}{4\kappa^2}(rS),$$

which is a standard second-order equation for the product rS with constant coefficients. Its general solution,[3] when written in a form that gives S explicitly, is

$$S = \frac{A}{r}\exp\left(-\frac{Ur}{2\kappa}\right) + \frac{B}{r}\exp\left(\frac{Ur}{2\kappa}\right).$$

Combining this solution with the definition of S given in (3.2) shows that

$$T = T_0 + \frac{A}{r}\exp\left\{\frac{U}{2\kappa}(x-r)\right\} + \frac{B}{r}\exp\left\{\frac{U}{2\kappa}(x+r)\right\}.$$

In most problems that occur in the theory of material processing, there is no additional heat input. The remainder of the workpiece acts as a heat sink and the value far from the origin is expected to tend to T_0. If we look at the exponential multiplying B, we see that $x + r > 0$ if $x > 0$, so that the coefficient of B tends to infinity as x tends to infinity, far downstream of the heat source. Since this is inconsistent with the normal conditions of such problems, the coefficient B must be zero.

It is now only necessary to provide an interpretation of A. A certain amount of care is needed at this point since the interpretation depends on whether the origin is at the surface of a workpiece or in the interior. The following discussion assumes that it is in the interior. Consider a small sphere Σ of radius a centered on the origin and estimate the flux of heat out of the surface of the sphere (see Figure 3.2). As much thermal

---

[3] Kreyszig, 1993, Ch. 2.

energy flows into the sphere across its boundaries as flows out across them. Hence, the only contribution to the flux is the part due to Fourier's law, and this has to be summed over the surface of the sphere; i.e., the total power flowing across the surface is

$$\int_{\Sigma}\left(-\lambda\frac{\partial T}{\partial r}\right)d\Sigma = -\lambda A \int_{\Sigma}\frac{\partial}{\partial r}\left\{\frac{1}{r}\exp\left[\frac{Ur}{2\kappa}(\cos\theta-1)\right]\right\}d\Sigma$$

in which the polar coordinate substitution $x = r\cos\theta$ has been used. It should be remembered that after the differentiation has been performed, r can be set equal to a.

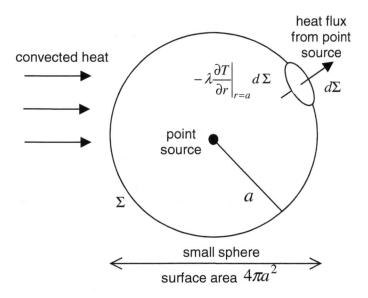

Figure 3.2. Flux of heat from a sphere of radius a centered on the origin.

After differentiation, the right-hand side of the equation is equal to

$$\lambda A \int_{\Sigma}\left\{\frac{1}{a^2}\exp\left[\frac{Ua}{2\kappa}(\cos\theta-1)\right]+O\left(\frac{1}{a}\right)\right\}d\Sigma .$$

The notation $O\left(\dfrac{1}{a}\right)$ means that these terms are only as big as $\dfrac{1}{a}$ if a is small, and can therefore be neglected compared to the other terms in the equation, as a is taken to be progressively smaller. For that reason, neglect these terms and notice, too, that the exponent is almost zero as well as a gets smaller. The integrand therefore becomes progressively more like $\dfrac{1}{a^2}$, which is a constant on the surface of Σ. The value of the integral is simply the surface area of the sphere, $4\pi a^2$, multiplied by $\dfrac{\lambda A}{a^2}$. Consequently, if the power supplied by the point source is P (W),

$$P = \frac{\lambda A}{a^2} \times 4\pi a^2 = 4\pi\lambda A$$

giving

$$A = \frac{P}{4\pi\lambda}.$$

The point source whose power is P in the interior of an unbounded workpiece, whose temperature far away is T_0, gives rise to a temperature distribution given by

$$T = T_0 + \frac{P}{4\pi\lambda r}\exp\left\{\frac{U}{2\kappa}(x-r)\right\} \qquad (3.4)$$

in which r is radial distance from the point source. Figure 3.3 shows the isotherms corresponding to such a point source embedded in an infinite medium. They are cylindrically symmetric about the x-axis. The diagram has been plotted in dimensionless form so that $T - T_0$ is scaled with $PU/k\kappa$, and the lengths x and r with $2\kappa/U$ so that

$$T = T_0 + \frac{PU}{2\kappa\lambda}\operatorname{point}\left(\frac{Ux}{2\kappa}, \frac{Uy}{2\kappa}, \frac{Uz}{2\kappa}\right)$$

with

$$\text{point}(x', y', z') = \frac{1}{4\pi} \frac{\exp\left(x' - \sqrt{x'^2 + y'^2 + z'^2}\right)}{\sqrt{x'^2 + y'^2 + z'^2}} \qquad (3.5)$$

and

$$x' = \frac{Ux}{2\kappa}, \; r' = \frac{Ur}{2\kappa}.$$

It can be helpful to define the function "point" in a way that is independent of the specific coordinate system in use; one way of doing so is to define

$$\text{point}(\mathbf{r}', \hat{\mathbf{u}}) = \frac{1}{4\pi} \frac{\exp(\mathbf{r}'.\hat{\mathbf{u}} - |\mathbf{r}'|)}{|\mathbf{r}'|}. \qquad (3.6)$$

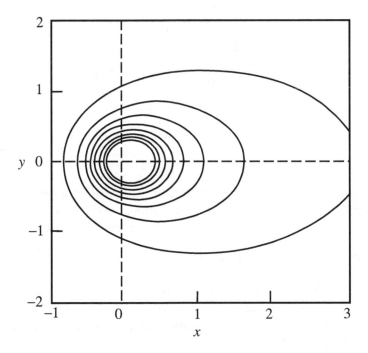

Figure 3.3. Contours of $\text{point}(x', 0, z')$; the contours are at intervals of 0.025 units in the range 0.025 to 0.2.

The second argument in this generalized notation for "point," written here as $\hat{\mathbf{u}}$, is a unit vector in the direction of translation and can usually be omitted since, in any given problem, it will always be the same.

There is a difference, however, if the point source is at the surface of a semi-infinite workpiece defined only in $z \geq 0$ (see Figure 3.4).

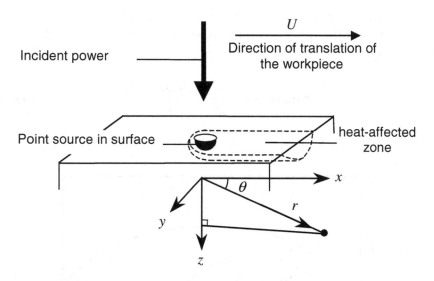

Figure 3.4. Point source on the surface of a workpiece.

The solution is still a valid temperature distribution so long as it is assumed that there is no heat loss by radiation, conduction, or any other means across the surface $z = 0$. By symmetry, the isotherms are perpendicular to the surface, thus ensuring that $-\lambda \dfrac{\partial T}{\partial z}\bigg|_{z=0} = 0$.

However, all the power now flows into the region $z \geq 0$, whereas before only half of it did, the rest flowing into $z < 0$. The departure of the temperature from T_0 is therefore double that given by (3.3), resulting in the temperature field

$$T = T_0 + \frac{P}{2\pi\lambda r}\exp\left\{\frac{U}{2\kappa}(x - r)\right\},$$

which can also be written

$$T = T_0 + \frac{PU}{\kappa\lambda}\,\text{point}(x', y', z')\tag{3.7}$$

for a point source of power P at the surface of a semi-infinite slab.

3.2.2 Applications of the point source solution

The point source solution can be very useful in the construction of solutions to problems in which there is a specified incident intensity distribution. The reason for this is because x, y, and z do not appear explicitly in the equation of heat conduction;

$$\text{point}\!\left(\frac{U}{2\kappa}(x - x_1), \frac{U}{2\kappa}(y - y_1), \frac{U}{2\kappa}(z - z_1)\right)$$

is also a solution of the equation, and corresponds to a point source of unit strength located at (x_1, y_1, z_1).

Suppose that the power absorbed is $I_a(\mathbf{r}_1)\delta S$ over an area δS centered on the point $\mathbf{r}_1 = (x_1, y_1, 0)$ in the surface of the workpiece. Then

$$\frac{I_a(\mathbf{r}_1)U}{\lambda\kappa}\,\text{point}\!\left(\frac{U(\mathbf{r} - \mathbf{r}_1)}{2\kappa}\right)\delta S$$

is the contribution to the temperature at point $\mathbf{r}$ from this particular element. The factor 2 comes from the fact that the power is absorbed in the surface of the workpiece, not its interior. See Equation (3.7). But in that case all that is needed is to use the linearity of the equation of heat conduction (when the coefficients are all constants) to add the individual contributions together for each such small element of area in the surface, giving a temperature distribution that is approximately

$$T = T_0 + \sum_{\text{all }\delta S} \frac{I_a(\mathbf{r}_1)U}{\lambda\kappa}\,\text{point}\!\left(\frac{U(\mathbf{r} - \mathbf{r}_1)}{2\kappa}\right)\delta S\ .$$

In the limit, as the diameter of every element in the surface tends to zero, this becomes an integral over the surface S of the workpiece, so that

$$T(\mathbf{r})=T_0 + \int_{\mathbf{r}_1 \in S}\frac{I_a(\mathbf{r}_1)U}{\lambda\kappa}\,\text{point}\!\left(\frac{U(\mathbf{r}-\mathbf{r}_1)}{2\kappa}\right)dS \qquad (3.8)$$

or

$$T(x,y,z)=T_0 +\frac{1}{2\pi\lambda}\int_{x_1=-\infty}^{\infty}\int_{y_1=-\infty}^{\infty} I_a(x_1,y_1)\times$$

$$\times\frac{\exp\!\left[\dfrac{U}{2\kappa}\!\left((x-x_1)-\sqrt{(x-x_1)^2+(y-y_1)^2+z^2}\right)\right]}{\sqrt{(x-x_1)^2+(y-y_1)^2+z^2}}\,dy_1 dx_1$$

$$(3.9)$$

in which $I_a(x,y)$ is the absorbed intensity distribution falling on the surface of the workpiece. Normally it is calculated from the relation

$$I_a(x,y)=(1-\mathcal{R})I(x,y).$$

I is the actual incident intensity and $\mathcal{R}$ is a suitably chosen reflection coefficient. An alternative way of expressing (3.9) is to make the substitution $x_1 = x - x_2, y_1 = y - y_2$. In that case, the alternative form of the integral is

$$T(x,y,z)=T_0 +\frac{1}{2\pi\lambda}\int_{x_2=-\infty}^{\infty}\int_{y_2=-\infty}^{\infty} I_a(x-x_2,y-y_2)\times$$

$$\times\frac{\exp\!\left[\dfrac{U}{2\kappa}\!\left(x_2-\sqrt{x_2^2+y_2^2+z^2}\right)\right]}{\sqrt{x_2^2+y_2^2+z^2}}\,dy_2 dx_2.$$

$$(3.10)$$

Either of these forms could be used with the following two cases as specially-shaped simple distributions of surface intensity.

The first is the case of a Gaussian beam of radius a centered on the origin for which

$$I(x, y) = \frac{P}{2\pi a^2} \exp\left(-\frac{x^2 + y^2}{2a^2}\right), \tag{3.11}$$

and the second is a simple model of a rectangular distribution of Gaussian beams.

(a)

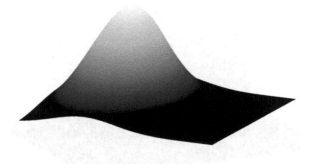

(b)

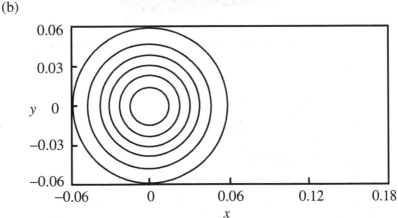

Figure 3.5. The Gaussian source is shown as a graph in (a), and the contours are shown in (b).

Suppose they are uniformly distributed over a rectangle centered at the origin, of sides $A \times B$ in the x and y directions, respectively. Each has the same power and has a beam radius a. If there are many individual beams in the array, the resulting intensity distribution is given approximately by[4]

$$I(x, y) = \frac{P}{4AB}\left\{\text{erf}\left(\frac{A - 2x}{2\sqrt{2}a}\right) + \text{erf}\left(\frac{A + 2x}{2\sqrt{2}a}\right)\right\}\left\{\text{erf}\left(\frac{B - 2y}{2\sqrt{2}a}\right) + \text{erf}\left(\frac{B + 2y}{2\sqrt{2}a}\right)\right\}.$$

(3.12)

(a)

(b)

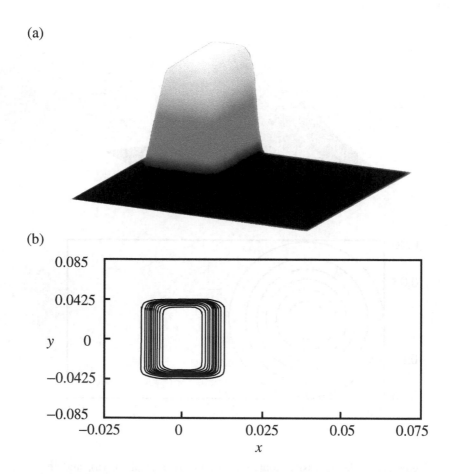

Figure 3.6. The rectangular array shown as (a) a graph and (b) as contours for a unit total power and $a = 0.0025$, $A = 0.02$, $B = 0.08$ units.

---

[4] For properties of the error function, erf, see Abramowitz and Stegun, 1965, Formula (7.1.1).

The total power in both cases is P. Figures 3.5 and 3.6 show graphs of these intensities together with the corresponding contour maps. Variation of the ratio $a:A:B$ affects the aspect of the fall-off region around the edge of the basic rectangle, with a small value of a giving a sharply-defined edge, and a larger value a less well-defined boundary. In general, the integrations will have to be performed numerically to find the temperature below the surface.

Equation (3.12) can be obtained from (3.11) by integrating it over the rectangle using the same power for each individual elementary Gaussian source. The same integration technique can be used to study the effect of arrays of other shapes.

Care must be taken, however, in the use of (3.10) if the incident intensity is discontinuous, as would be the case of the "top-hat" intensity distribution

$$I(x, y) = \begin{cases} P/\pi a^2 & \sqrt{x^2 + y^2} < a \\ 0 & \text{otherwise.} \end{cases} \tag{3.13}$$

A polar coordinate transformation may sometimes be useful in the evaluation of these integrals. The region of integration for (3.10) defined by $-\infty < x_2 < \infty$, $-\infty < y_2 < \infty$, for example, can be described equally well in terms of the polar co-ordinates (r, θ) where $x_2 = r \cos \theta$, $y_2 = r \sin \theta$ with $-\pi < r \leq \pi, 0 \leq r < \infty$. The natural elements of area for the integration, however, are no longer small rectangles of area $dx_2 dy_2$ with sides parallel to the x_2, y_2 coordinate axes. They are now almost rectangular elements whose sides are $r d\theta$ and dr in the directions of increasing θ and increasing r, respectively (see Figure 3.7). The magnitude of the element of area is therefore $r \, dr \, d\theta$. In that case, Equation (3.10) for the temperature distribution becomes

$$T(x, y, z) = T_0 + \frac{1}{2\pi\lambda} \int_{r=0}^{\infty} \int_{\theta=-\pi}^{\pi} I_a(x - r \cos \theta, y - r \sin \theta) \times$$

$$\times \frac{\exp\left[\dfrac{U}{2\kappa}\left(r \cos \theta - \sqrt{r^2 + z^2}\right)\right]}{\sqrt{r^2 + z^2}} r d\theta \, dr$$

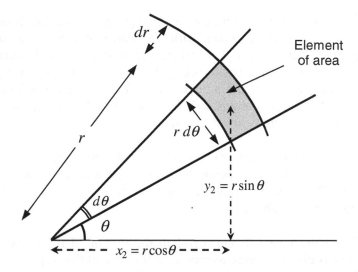

Figure 3.7. The element of area in a polar coordinate system.

and a double numerical integration is usually necessary. There is one circumstance where a worthwhile approximation is sometimes available. Suppose that the intensity distribution is characterized by a length scale a so that outside a distance from the origin of this order, I_a is effectively zero. In that case, there is a Péclet number $\text{Pe} = Ua/2\kappa$, and this determines the relative dominance of the exponential term $\exp\left[\dfrac{U}{2\kappa}\left(r\cos\theta - \sqrt{r^2 + z^2}\right)\right]$. In the plane of integration this has its greatest value for any given r when $\theta = 0$ when it has the value $\exp\left[\dfrac{U}{2\kappa}\left(r - \sqrt{r^2 + z^2}\right)\right]$.

If the Péclet number is large, the exponential falls very rapidly from the maximum value as θ departs from zero and the integrand is effectively zero. What this means is that if $\cos\theta$ is replaced by its local approximation, $1 - \tfrac{1}{2}\theta^2$, the error introduced will be very small indeed. Furthermore, the range of integration can even be extended (purely for convenience) to $-\infty < \theta < \infty$ without significant additional error. The contribution from the intensity can then be approximated by the value local to $\theta = 0$. It does not vary rapidly on the length-scale a, unlike the exponential, so it will not have changed significantly by the time the exponential has become very small.

(a)

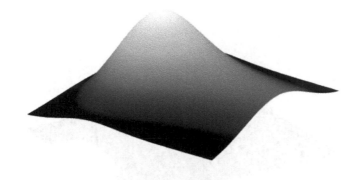

(b)

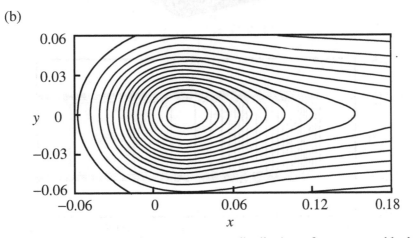

Figure 3.8. The surface temperature distribution of a concrete block, calculated for a Gaussian beam with $a = 3\,\text{cm}$, $P = 1.9\,\text{kW}$, and $U = 10\,\text{cm s}^{-1}$. The surface temperature on the axis of the laser beam is, in this case, predicted to be 331 K above ambient. Shown as (a) a graph and (b) a contour diagram.

The expression for the temperature can thus be approximated by

$$T(x, y, z) = T_0 + \frac{1}{2\pi\lambda} \int_{r=0}^{\infty} I_a(x-r, y) \frac{\exp\left[\frac{U}{2\kappa}\left(r - \sqrt{r^2 + z^2}\right)\right]}{\sqrt{r^2 + z^2}} \times$$

$$\times \left[\int_{\theta=-\infty}^{\infty} \exp\left(-\frac{Ur}{4\kappa}\theta^2\right) d\theta\right] r \, dr$$

(a)

(b)

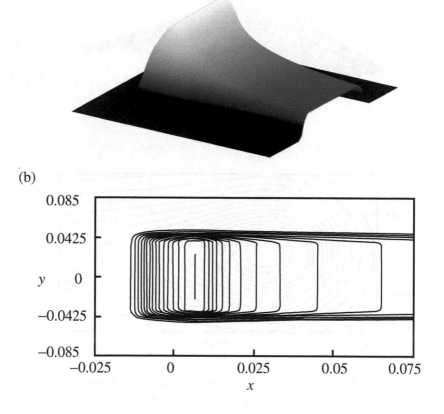

Figure 3.9. The incident laser intensity and the resulting surface temperature distribution of a concrete block, calculated for a quasi-rectangular beam with $A = 2$ cm, $B = 8$ cm, $a = 2.5$ mm, $P = 1.9$ kW, and $U = 10$ cm s^{-1}. The surface temperature on the axis of the laser beam is then predicted as 621 K above ambient. Shown as (a) a graph and (b) a contour diagram.

But[5]

$$\int_{\theta=-\infty}^{\infty} \exp\left(-\frac{Ur}{4\kappa}\theta^2\right) d\theta = 2\sqrt{\frac{\kappa\pi}{Ur}}$$

so that

$$T(x,y,z) = T_0 + \frac{1}{\lambda}\sqrt{\frac{\kappa}{\pi U}} \int_{r=0}^{\infty} I_a(x-r,y)\sqrt{\frac{r}{r^2+z^2}}\, e^{\frac{U}{2\kappa}\left(r-\sqrt{r^2+z^2}\right)}\, dr \,.$$

$$(3.14)$$

The advantage here is that only a single integral needs to be evaluated rather than a double one. Figure 3.8 shows a graph of the surface temperature that results from a Gaussian surface intensity distribution (3.11), while Figure 3.9 shows the graph of temperature at the surface from the quasi-rectangular distribution given by (3.12). In both cases the parameters chosen for the purpose of the numerical examples are those typical for the laser scabbling of concrete; qualitatively, however, the results are typical when the Péclet number is high.

From (3.14) it follows that when the Péclet number is high, the order of magnitude of the temperature rise is

$$\frac{P}{\lambda}\sqrt{\frac{\kappa}{a^3 U}} \,.$$

$$(3.15)$$

In the case of concrete, for example, it leads to a maximum temperature of 331 K above ambient in the case of the given Gaussian beam, and 621 K in the case of the quasi-rectangular beam.

3.3 THE TEMPERATURE DISTRIBUTION IN PLATES

The techniques discussed so far for finding the temperature distribution have assumed that there is a heat input at the surface $z = 0$ of a semi-infinite workpiece that occupies the region $z \geq 0$. It has also

---

[5] Kreyszig, 1993, Sec. 9.3.

been assumed that there is no heat loss from the surface, either by evaporation from any region of liquid metal (if the temperature is high enough for there to be one), by radiation, by convective cooling, or, indeed, by any other mechanism. In that case, the surface condition on the temperature is simply that

$$\frac{\partial T}{\partial z} = 0,\tag{3.16}$$

except on those parts of the surface where there is a known power input, in which case the z derivative of T is known. In most of the problems considered, most of the mechanisms neglected result in only very small heat losses. The one exception is the case of convective heat loss, which can sometimes be significant. A simple model is to assume that it is proportional to the excess temperature of the surface. This contribution is very often neglected when it is not the principal subject of the investigation for which the model is being constructed, but in practice it can be significant.

The solutions considered so far assume that the workpiece is very thick – thick enough, that is, for the effect on the temperature distribution of the lower boundary to be negligible. From the solution for the point source in the form given in Equation (3.5) it is clear that there is a length scale in the problem equal to $2\kappa/U$. This is in addition to the length scale, a, of the incident intensity and of the thickness of the workpiece itself, h. From the expression for the temperature in the interior of the workpiece given by the convolution in Equation (3.8) it follows that the temperature tends to zero in a manner that depends primarily on that length scale. Consequently, if

$$h \gg \frac{2\kappa}{U},\tag{3.17}$$

the presence of the bottom boundary does not play a significant role. If, however, the condition is not satisfied some account must be taken of it.

In the same way that heat loss at the surface has been ignored, only the case where there is no heat loss from the bottom will be considered so that condition (3.16) also applies at $z = h$. If a solution $S(x, y, z)$

has been found for the case of the infinite slab and it satisfies the conditions

$$S(x, y, z) \text{ is a solution of (2.67) with } q = 0;$$

S is an even function in z (so its z derivative is odd);

$$S_z(x, y, 0) = -I(x, y)/\lambda;$$

$$S \to 0 \text{ as } z \to \pm\infty.$$

I is the known incident intensity (and thus includes the condition of no heat loss from those parts of the surface not subject to heating) and the subscript z indicates differentiation with respect to z, then consider

$$S(x, y, z) + S(x, y, z - 2h),$$

which represents an addition of the same solution, but with its origin shifted to $z = 2h$. Since its z-derivative is $S_z(x, y, z) + S_z(x, y, z - 2h)$ it satisfies the condition that its z-derivative vanishes at $z = h$, since $S_z(x, y, -h) = -S_z(x, y, h)$. Unfortunately, it no longer satisfies the condition at $z = 0$. That, however, can be put right by adding on another term, $S(x, y, z + 2h)$, equivalent to adding on the same solution with its origin shifted to $z = -2h$. The z-derivative of the sum of these three terms is $S_z(x, y, z) + S_z(x, y, z - 2h) + S_z(x, y, z + 2h)$, which at $z = 0$ now has the value $S_z(x, y, 0) - S_z(x, y, 2h) + S_z(x, y, 2h)$. This is equal to $-I/k$. The condition at $z = h$ is no longer right, however. It can be corrected by yet another addition, and so on. Because S tends to zero as its z-argument tends to plus or minus infinity, the effect of these additional terms gets progressively less. The result is a sequence of increasingly good approximations to the solution of the problem for the slab of finite thickness.

The process can be regarded as taking the original solution, and adding on the same solution with its origin shifted alternately to reflections in $z = h$ and in $z = 0$, as illustrated in Figure 3.10. The process leads to the final expression for the temperature distribution in the plate $0 \le z \le h$ given by

$$T(x, y, z) = T_0 + S(x, y, z) + \sum_{n=1}^{\infty} S(x, y, z - 2nh) + \sum_{n=1}^{\infty} S(x, y, z + 2nh),$$

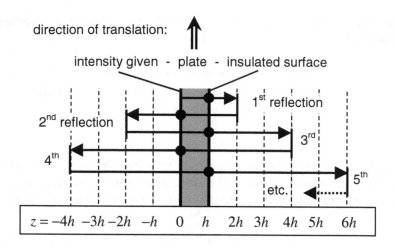

Figure 3.10. The method of images: the way in which the image system is built up by progressive shifts of the origin of *S*.

from which it is easy to verify that the two boundary conditions $-\lambda T_z(x,y,0)=I(x,y)$ and $T_z(x,y,h)=0$ are satisfied, remembering that S_z is odd in its third argument. A simpler way to write the solution is

$$T(x,y,z)=T_0 + \sum_{n=-\infty}^{\infty} S(x,y,z-2nh).\qquad(3.18)$$

Because of the way in which the solution is built up by means of reflections, it is often known as the "method of images",[6] especially when used in conjunction with the point and line source solutions. It can be generalized to other boundary shapes.

For example, the temperature distribution in a semi-infinite work-piece given by a point source at the surface is given by Equation (3.7) as

$$T'=\frac{1}{2\pi}\frac{\exp\left\{x'-\sqrt{x'^2+y'^2+z'^2}\right\}}{\sqrt{x'^2+y'^2+z'^2}},$$

---

[6] Carslaw and Jaeger, 1959, 273-281.

where the variables have all been made dimensionless and the temperature is referred to the ambient value. Figure 3.11 shows the temperature contours for this solution in $z' > 0$, in the plane of symmetry $y' = 0$ with the level corresponding to $z' = 1$, indicated by a broken line. The solution constructed from this by the method of images using Equation (3.18), for a plate whose bottom surface is at $z' = h'$, is

$$T' = \frac{1}{2\pi} \sum_{n=-\infty}^{\infty} \frac{\exp\left\{x' - \sqrt{x'^2 + y'^2 + (z' - 2nh')^2}\right\}}{\sqrt{x'^2 + y'^2 + (z' - 2nh')^2}} .$$

The corresponding set of contours in $y = 0$ is drawn in Figure 3.12 for the case when $h' = 1$. Notice the way that the isotherms in 3.12 are perpendicular to the bottom of the workpiece at $z' = h'$.

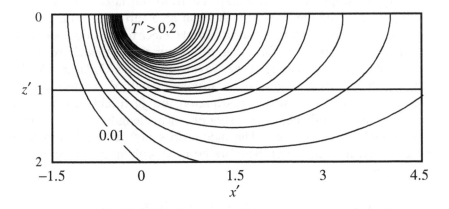

Figure 3.11. Isotherms in the plane of symmetry for the point source solution; the line given by $z' = 1$ is shown for comparison with the isotherms in Figure 3.12.

Since the point source solution satisfies the conditions necessary for the application of the method of images, a solution constructed from the point source solution by the method of Section 3.2 also satisfies the conditions.

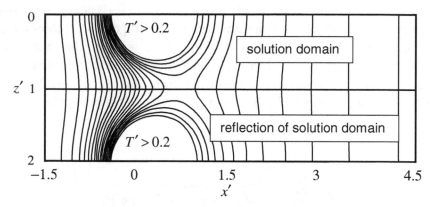

Figure 3.12. Isotherms in the plane of symmetry for a workpiece of dimensionless thickness $h' = 1$ constructed using the method of images.

3.4 FRESNEL ABSORPTION

Energy absorption by the workpiece from the laser can involve a direct process[7] with the laser light incident on a surface as well as other indirect processes. There exists a simple electromagnetic model for absorption at a metal surface that is widely used, especially in the context of keyhole modeling at the wavelength of a CO_2 laser. At that wavelength the assumptions of a simple model of electromagnetic interaction involving resistive dissipation are justifiable as a useful approximation, although at shorter wavelengths it becomes progressively more suspect. The model does not make allowance for surface impurities and must therefore be used with an understanding of its limitations.

This direct absorption process is usually referred to as *Fresnel absorption*. A formula for the reflection coefficient $\mathcal{R}$ that is frequently quoted,[8] and which applies to circularly polarized light, is

$$\mathcal{R} = \frac{1}{2}\left(\frac{1 + (1 - \varepsilon\cos\theta)^2}{1 + (1 + \varepsilon\cos\theta)^2} + \frac{\cos^2\theta + (\varepsilon - \cos\theta)^2}{\cos^2\theta + (\varepsilon + \cos\theta)^2} \right) \qquad (3.19)$$

[7] Pirri et al., 1978; Schulz et al., 1986; Schellhorn and Spindler, 1987; Solana et al., 2000.
[8] Stratton, 1941, 500-11; Schulz et al., 1987.

where θ is the angle of reflection that the light makes to the normal and ε is a material-dependent quantity defined by

$$\varepsilon^2 = \frac{2\varepsilon_2}{\varepsilon_1 + \sqrt{\varepsilon_1^2 + (\sigma_{st} / \omega\varepsilon_0)^2}}$$

where ε_0 is the permittivity of a vacuum, ε_1 and ε_2 are the real parts of the dielectric constants for the metal and the air or vapor through which the beam is being transmitted, and σ_{st} is the electrical conductance per unit depth of the workpiece. The value of ε_0 is 8.854×10^{-12} F m^{-1}, and typical values for the other terms are approximately unity for ε_1 and ε_2, and 5.0×10^5 Ω^{-1}m^{-1} for σ_{st}. For a CO_2 laser a wavelength of $10.6\,\mu$m gives a value for ω of 1.78×10^{14} s^{-1}, and so ε has a value of about 0.08. Figure 3.13 shows the graph of $\mathcal{R}$ as a function of θ.

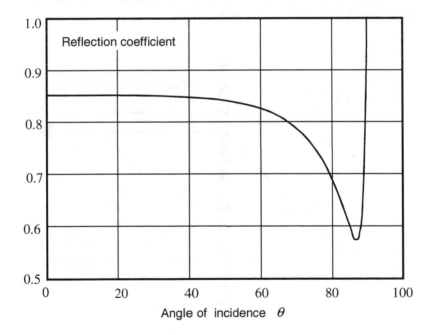

Figure 3.13. The reflection coefficient $\mathcal{R}$ as a function of the angle of the incident beam to the normal.

It will be seen that there is a strong dependence on the angle of incidence with a marked minimum close to $\theta = \frac{1}{2}\pi$, indicating that absorption is strongest at near-grazing incidence. For normal incidence, however, as much as 85% of the incident power can be reflected. This figure can be very greatly modified by surface impurities or other additives introduced as part of the process.

3.5 THE LINE SOURCE SOLUTION

The line source solution can be obtained in essentially the same way as the point source. The difference is that a solution is sought that depends only on the coordinate in the direction of translation x and distance from the z-axis, but is independent of distance in the z-direction. As before, it is assumed that the workpiece is moving steadily parallel to the x-axis with velocity U, and that all the material parameters are constants. Consequently, the temperature satisfies Equation (3.1); Figure 3.14 shows the relative geometry and the co-ordinate system. Once again, look for a solution of the form given by Equation (3.2) so that S is given by (3.3).

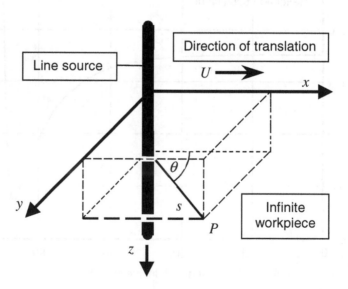

Figure 3.14. The relative geometry and the coordinate system employed in the description of the line source solution. The point P is given by $x = s\cos\theta$, $y = s\sin\theta$, z.

This time, however, look for a solution for S that depends only on distance s from the z-axis, $s = \sqrt{x^2 + y^2}$ so that[9]

$$\frac{\partial S}{\partial x} = \frac{\partial s}{\partial x}\frac{dS}{ds} = \frac{x}{s}\frac{dS}{ds}.$$

Consequently,

$$\frac{\partial^2 S}{\partial x^2} = \frac{1}{s}\frac{dS}{ds} + \frac{x^2}{s^2}\frac{d^2 S}{ds^2} - \frac{x^2}{s^3}\frac{dS}{ds}$$

and similarly for the second derivative of S with respect to z. The second derivative with respect to z is unaltered in form. Adding them together shows that

$$\frac{d^2 S}{ds^2} + \frac{1}{s}\frac{dS}{ds} - \frac{U^2}{4\kappa^2}S = 0.$$

Consultation of any standard reference work on the mathematics of physics or differential equations[10] shows that the general solution of this equation is

$$S = A\, I_0\!\left(\frac{Us}{2\kappa}\right) + B\, K_0\!\left(\frac{Us}{2\kappa}\right)$$

where I_0 and K_0 are the modified Bessel functions of order 0.

In order to make use of this solution it is helpful to understand the character of the two functions I_0 and K_0. Figure 3.15 shows the graphs of each of them on the same axis. The most important points to notice are the following.
- I_0 increases indefinitely as s' increases;
- K_0 tends to zero as s' increases;

---

[9] It should be remembered that the use of r to mean both radial distance from the origin of spherical polar coordinates and the distance from the axis in cylindrical polars is very common. The symbol s is used here to avoid confusion, but the usage is not standard.

[10] Abramowitz and Stegun, 1965, 9.6.1.

- K_0 is unbounded (tends to infinity) as s' tends to zero.

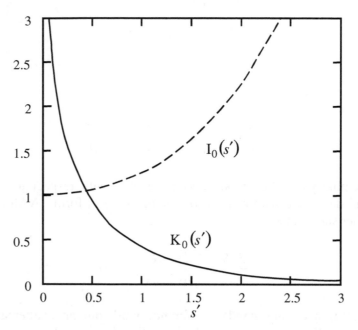

Figure 3.15. The functions $I_0(s')$ (broken line) and $K_0(s')$ (solid line).

If this solution for S is combined with the definition of S given in (3.2), it follows that

$$T = T_0 + A \exp\left(\frac{U}{2\kappa}x\right) I_0\left(\frac{Us}{2\kappa}\right) + B \exp\left(\frac{U}{2\kappa}x\right) K_0\left(\frac{Us}{2\kappa}\right).$$

The condition that there is no additional heat input to the remainder of the workpiece, which acts as a heat sink, once again imposes a restriction on the coefficients A and B. For large values of s the asymptotic forms of the Bessel functions[11] show that

$$T \sim T_0 + A\sqrt{\frac{\kappa}{\pi Us}} \exp\left\{\frac{U}{2\kappa}(x+s)\right\} + B\sqrt{\frac{\pi\kappa}{Us}} \exp\left\{\frac{U}{2\kappa}(x-s)\right\}.$$

---

[11] Abramowitz and Stegun, 1965, 9.7.1 and 9.7.2.

Looking at the multiple of A, notice that $x+s>0$ if $x>0$, so that the coefficient of A tends to infinity as x tends to infinity far from the heat source; consequently, the coefficient A must be zero.

An interpretation of B is possible in the same way as before; once again some care is needed depending on whether the z-axis is in the surface of a semi-infinite workpiece or in the interior of an infinite one. The following discussion assumes that it is in the interior.

Consider a small cylinder C of radius a centered on the axis and estimate the flux of heat from its surface per unit length; see Figure 3.16. As much thermal energy flows into the cylinder across its boundaries as flows out across them. The only contribution to the flux is the part due to Fourier's law, and this has to be summed over the surface of the cylinder; i.e., the total power flowing across the surface per unit length is

$$\int_C \left(-\lambda \frac{\partial T}{\partial s} \right) dC = -\lambda B \int_C \frac{\partial}{\partial s} \left\{ \exp\left(\frac{U}{2\kappa} s\cos\theta \right) K_0\left(\frac{Us}{2\kappa} \right) \right\} dC \quad (3.20)$$

where the polar coordinate substitution $x = s\cos\theta$ has been used. It should be remembered that after the differentiation has been performed, s can be set equal to a.

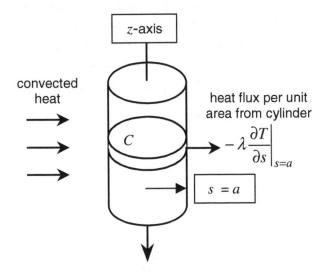

Figure 3.16. Flux of heat out of a cylinder of radius a centered on the z-axis.

Now [12]

$$K_0(s') = -\left\{\ln\tfrac{1}{2}s' + \gamma\right\}0(s'^2)$$

where $\gamma = 0.57721566...$ is the Euler-Mascheroni constant, so that

$$\frac{\partial}{\partial s}\left\{\exp\left(\frac{U}{2\kappa}s\cos\theta\right)K_0\left(\frac{Us}{2\kappa}\right)\right\} =$$

$$= \frac{\partial}{\partial s}\left\{\exp\left(\frac{U}{2\kappa}s\cos\theta\right)\left[-\ln\tfrac{1}{2}s - \gamma + 0(s^2)\right]\right\} = -\frac{1}{s} + O(\ln s).$$

The integral of the right-hand side of Equation (3.20) is consequently equal to

$$-\lambda B\int_C\left\{-\frac{1}{a} + O(\ln a)\right\}dC.$$

The notation $O(X)$ means that these terms are only as big as X. For that reason, neglect these terms and notice that the exponent is almost zero as well as a gets smaller. The integrand becomes progressively more like $1/a$ since $a\ln a \to 0$ as $a \to 0$, and is therefore constant on the surface of C for sufficiently small a. Consequently, the value of the integral is just the length of the circumference of the circle of cross-section, $2\pi a$, multiplied by $\frac{\lambda B}{a}$. Hence, if the power supplied by the line source per unit length is Q,

$$Q = \frac{\lambda B}{a} \times 2\pi a = 2\pi\lambda B,$$

giving

$$B = \frac{Q}{2\pi\lambda}.$$

---

[12] Abramowitz and Stegun, 1965, 9.6.13 with 9.6.12.

As a result, the line source whose power is Q (W m^{-1}) in the interior of an unbounded workpiece whose temperature far away is T_0 gives rise to a temperature distribution given by

$$T = T_0 + \frac{Q}{2\pi\lambda} \exp\left(\frac{Ux}{2\kappa}\right) K_0\left(\frac{Us}{2\kappa}\right) \tag{3.21}$$

where s is perpendicular distance from the line source.

Figure 3.17 shows the isotherms corresponding to such a line source embedded in an infinite medium, and Figure 3.18 shows the same thing as a graph. The diagram has been plotted in dimensionless form so that $T - T_0$ is scaled with $QU/\lambda\kappa$ and the lengths x and r with $2\kappa/U$ so that

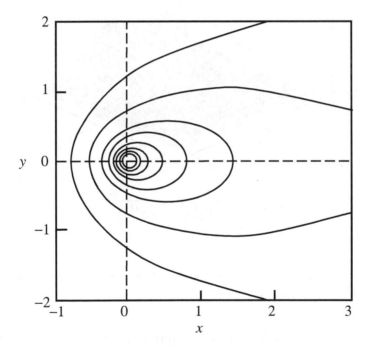

Figure 3.17. Contours of $\dfrac{\exp(x)}{2\pi} K_0\left(\sqrt{x^2 + y^2}\right)$; contours are at intervals of 0.05 units in the range 0.05 to 0.4.

$$T = T_0 + \frac{Q}{\lambda} \text{line}(x', y')$$

where

$$\text{line}(x', z') = \frac{1}{2\pi} \exp(x') K_0\left(\sqrt{x'^2 + y'^2}\right) \qquad (3.22)$$

with

$$x' = \frac{U x}{2\kappa}, \quad y' = \frac{U y}{2\kappa}.$$

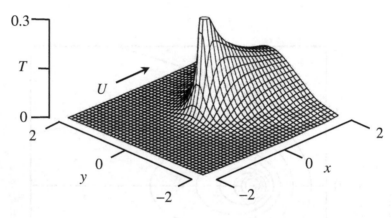

Figure 3.18. Graph of $\dfrac{\exp(x')}{2\pi} K_0\left(\sqrt{x'^2 + y'^2}\right)$.

A coordinate-free definition can again be useful, in which case the function "line" can be defined by

$$\text{line}(s', \hat{\mathbf{u}}) = \frac{1}{2\pi} \exp(s'.\hat{\mathbf{u}}) K_0(s'), \qquad (3.23)$$

in which the second argument of line, written here as $\hat{\mathbf{u}}$, is a unit vector in the direction of translation and can often be left out. The vector $\mathbf{s}'$ is the vector perpendicular to the line source to the point P in space given by the position vector $\mathbf{r}'$ (see Figure 3.14). It is related to $\mathbf{r}'$ by $\mathbf{s}' = \mathbf{r}' - (\mathbf{r}'.\mathbf{k})\mathbf{k}$ where $\mathbf{k}$ is the axis of the line source and has to be perpendicular to the direction of translation, so that $\hat{\mathbf{u}}.\mathbf{k} = 0$.

There is a difference if the line source lies along the y-axis in the surface of a semi-infinite workpiece defined only in $z \geq 0$ (see Figure 3.19).

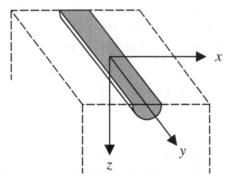

Figure 3.19. Line source on the surface of a workpiece.

As before, the symmetry of the isotherms ensures that the solution is still a valid temperature distribution so long as no account is taken of heat loss by radiation, convective cooling, or any other means across the surface $z = 0$. The isotherms are perpendicular to the surface, thus ensuring that $-\lambda \dfrac{\partial T}{\partial z}\bigg|_{z=0} = 0$, and so the solution is compatible with the assumption that there is no heat loss across the boundary. However, all the power now flows into the region $z \geq 0$, whereas before only half of it did, the rest flowing into $z < 0$. The departure of the temperature from T_0 is therefore double that given by (3.21), resulting in the temperature distribution

$$T = T_0 + \frac{Q}{\pi\lambda}\exp\left(\frac{Ux}{2\kappa}\right)K_0\left(\frac{U}{2\kappa}\sqrt{x^2 + z^2}\right)$$

$$= T_0 + 2\frac{Q}{\lambda}\mathrm{line}(x', z')$$

(3.24)

for a line source of strength Q (W m^{-1}) on the surface of a semi-infinite slab.

Just as the point source solution can be used to construct further solutions by adding together individual solutions or, in the limit, integrating over them, so, too, the line source solution can be used in exactly the same way. Suppose it is desired to find the temperature distribution in a semi-infinite block whose surface $z = 0$ is subject to an incident intensity, whose absorbed value is $I_a(x)$ W m^{-2}, and is independent of the lateral coordinate y. If we consider just the effect due to absorption in a strip of width δx_1 in the surface at $x = x_1$, the rise in temperature at the point (x, z) caused by absorption in the strip is

$$\delta T = 2 \frac{I_a(x_1)\delta x_1}{\lambda} \text{line}\left(\frac{U(x - x_1)}{2\kappa}, \frac{Uz}{2\kappa}\right).$$

To find the rise in temperature caused by all of these small strips it is only necessary to add them all together or, in the limit, to use integration. The result is that

$$T = T_0 + \frac{2}{\lambda} \int_{x_1 = -\infty}^{\infty} I_a(x_1) \text{line}\left(\frac{U(x - x_1)}{2\kappa}, \frac{Uz}{2\kappa}\right) dx_1. \qquad (3.25)$$

Alternatively, the substitution $x_1 = x - x_2$ can be made so that the temperature distribution is given by

$$T = T_0 + \frac{2}{\lambda} \int_{x_2 = -\infty}^{\infty} I_a(x - x_2) \text{line}\left(\frac{Ux_2}{2\kappa}, \frac{Uz}{2\kappa}\right) dx_2. \qquad (3.26)$$

These formulae can be applied for example to the one-dimensional Gaussian distribution

$$I(x) = Q \frac{\exp(-x^2/2a^2)}{a\sqrt{2\pi}} \qquad (3.27)$$

and the one-dimensional "top-hat" distribution

$$I(x) = \frac{Q}{2a} H(a - |x|) = \begin{cases} 0 & x < -a \\ Q/2a & -a < x < a \\ 0 & x > a \end{cases} \qquad (3.28)$$

in which $H(x)$ is the Heaviside step function whose value is zero when its argument is negative, and 1 when it is positive. In the problems considered, its value at the point of discontinuity is irrelevant provided, in practice, the intensity distribution is not characterized by a sharp spike at these points. In each case Q is the absorbed power per unit width of the workpiece and $2a$ is the beam width. Either formula can be used, but in the case of the top-hat distribution, care is needed with (3.26), as the integral is over a finite range from $x - a$ to $x + a$.

If Equation (3.26) is used, the problem with the top-hat distribution has a solution that can be written[13]

$$T = T_0 + \frac{Q}{2\pi\lambda a} \int_{x_1 = x-a}^{x+a} \exp\left(\frac{Ux_1}{2\kappa}\right) K_0\left(\frac{U}{2\kappa}\sqrt{x_1^2 + z^2}\right) dx_1;$$

it can also be written as

$$T = T_0 + \frac{Q}{2\pi\lambda} f\left(\frac{Ux}{2\kappa}, \frac{Uz}{2\kappa}, \frac{Ua}{2\kappa}\right) \qquad (3.29)$$

with

$$f(x', z', \mathrm{Pe}) = \frac{1}{\mathrm{Pe}} \int_{x_1' = x - \mathrm{Pe}}^{x' + \mathrm{Pe}} \exp x_1' \, K_0\left(\sqrt{x_1'^2 + z'^2}\right) dx_1'. \qquad (3.30)$$

The form for f is obtained by means of the substitutions $x' = Ux/2\kappa$, $x_1' = Ux_1/2\kappa$, $z' = Uz/2\kappa$ and $\mathrm{Pe} = Ua/2\kappa$. Contours of f for a Péclet number Pe equal to 1 are shown in Figure 3.20, and for Pe = 10 in Figure 3.21. Note the way in which the maximum temperature and the

---

[13] Carslaw and Jaeger, 1959, 268. Note that the solution given there is for a line source moving through the interior of an infinite medium, and their symbol Q is the quantity used here divided by $2a$; a is b in their notation.

thickness of the layer affected both decrease as the Péclet number increases. The maximum temperature always occurs on the surface.

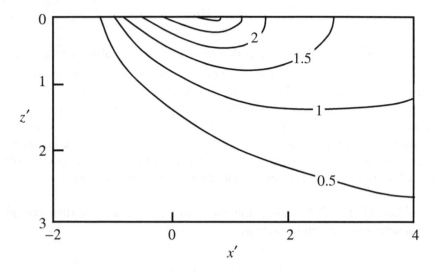

Figure 3.20. Contours of the dimensionless temperature $f(x', z', \mathrm{Pe})$ for the heated strip for a Péclet number of 1. Contours are at intervals of 0.5.

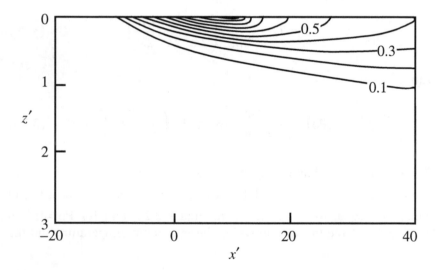

Figure 3.21. Contours of the dimensionless temperature $f(x', z', \mathrm{Pe})$ for the heated strip for a Péclet number of 10. Contours are at intervals of 0.1.

The integral in Definition (3.30) normally has to be evaluated numerically, but if the point in question is on the surface so that $z' = 0$ it becomes

$$f(x',z',\text{Pe}) = \frac{1}{\text{Pe}} \int_{x_1'=x'-\text{Pe}}^{x_1'=x'+\text{Pe}} \exp x_1' K_0 |x_1'| dx_1' ,$$

an expression that can be evaluated explicitly using King's integral,[14]

$$\int_{t=0}^{z} e^{\pm t} K_0(t) dt = z e^{\pm z} \{K_0(z) \pm K_1(z)\} \mp 1, \quad z > 0.$$

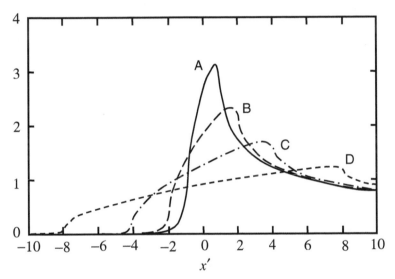

Figure 3.22. Graph of the temperature at the surface for different values of the Péclet number, Pe. A: Pe = 1; B: Pe = 2; C: Pe = 4; D: Pe = 8.

The result is that

$$f(x',0,\text{Pe}) =$$
$$\frac{1}{\text{Pe}} \{(x' + \text{Pe})\exp(x' + \text{Pe})[K_0|x' + \text{Pe}| + \text{sign}(x' + \text{Pe})K_1|x' + \text{Pe}|]\} -$$
$$-\frac{1}{\text{Pe}} \{(x' - \text{Pe})\exp(x' - \text{Pe})[K_0|x' - \text{Pe}| + \text{sign}(x' - \text{Pe})K_1|x' - \text{Pe}|]\}$$

$$(3.31)$$

---

[14] Abramowitz and Stegun, 1965, Formula 11.3.16.

in which sign(t) is +1 if t is positive, and −1 if t is negative.

Expression (3.31) can be used to calculate the temperature at the surface for different values of the Péclet number. Graphs of the results are shown in Figure 3.22. Note that the maximum value falls as the Péclet number increases. The location of the maximum remains near to the edge of the strip on which the incident intensity falls. This is at $x' = 1, 2, 4, 8$ in the four cases shown, corresponding to $Pe = 1, 2, 4, 8$, since the strip occupies $-Pe < x' < Pe$.

CHAPTER 4

TIME-DEPENDENT SOLUTIONS IN BLOCKS

4.1 TIME-DEPENDENT ONE-DIMENSIONAL SOLUTIONS

Consider the problem of the semi-infinite rod in $x > 0$ that is initially at temperature T_0 everywhere along its length. At time $t = 0$ the end at $x = 0$ is raised to temperature T_1. A solution can be found using dimensional analysis because there are so few dimensional quantities involved in the problem. A temperature scale is provided by the difference $T_1 - T_0$. The only other dimensional quantities in the problem are the thermal diffusivity κ (units, m^2 s^{-1}), the length x, and the time t. Only one independent dimensionless ratio can be formed from these, namely $x^2/\kappa t$. This can be seen by considering $\kappa^a x^b t^c$, whose units will be m^{2a+b} s^{-a+c}. These will be dimensionless if $2a + b = -a + c = 0$ leading to the ratio $a:b:c = 2:-1:-1$. Any function of these will also be dimensionless, and it proves to be somewhat more convenient if the combination

$$\eta = x/\sqrt{4\kappa t}$$

is used instead. These considerations suggest that the one-dimensional time-dependent conduction equation,

$$\frac{\partial T}{\partial t} = \kappa \frac{\partial^2 T}{\partial x^2}, \tag{4.1}$$

has a solution of the following form,

$$T = T_0 + (T_1 - T_0)f(\eta).$$

Substitution of such a form into Equation (4.1) shows that f must satisfy the equation

$$\frac{d^2 f}{d\eta^2} = -2\eta \frac{df}{d\eta}.$$

(4.2)

The initial condition at $t=0$ and the asymptotic condition as $x \to \infty$ both require that $f \to 0$, as $\eta \to \infty$ and the boundary condition $T(0,t)=T_1$ require that $f(0)=1$. Equation (4.2) can be regarded as a first-order equation in $df/d\eta$, which can be solved by separating its variables[1] so its first integral is

$$\frac{df}{d\eta} = A e^{-\eta^2}.$$

The solution that satisfies $f(0)=1$, is therefore,

$$f = 1 + A \int_{s=0}^{\eta} e^{-s^2}\, ds.$$

The requirement that $f \to 0$ as $\eta \to \infty$ means that

$$A = 1 \Big/ \int_{s=0}^{\infty} e^{-s^2}\, ds = \frac{2}{\sqrt{\pi}}.$$

The error function and the complementary error function[2] are defined by

$$\mathrm{erf}(x) = \frac{2}{\sqrt{\pi}} \int_{s=0}^{x} e^{-s^2}\, ds \ \text{ and } \ \mathrm{erfc}(x) = 1 - \mathrm{erf}(x) = \frac{2}{\sqrt{\pi}} \int_{s=x}^{\infty} e^{-s^2}\, ds.[3]$$

The solution for the problem under consideration is

$$T = T_0 + (T_1 - T_0)\,\mathrm{erfc}\!\left(\frac{x}{\sqrt{4\kappa t}}\right).$$

(4.3)

---

[1] Kreyszig, 1993, Ch.1.
[2] Abramowitz and Stegun, 1965, Equations (7.1.1) and (7.1.2).
[3] Kreyszig, 1993, Appendix 3.

The graph of this solution as a function of x is shown in dimensionless form in Figure 4.1 for a number of different values of $\sqrt{4\kappa t}$. It has been drawn in terms of an arbitrary length scale a, which can be whatever is most convenient, 1 mm, 10 cm, or any other scale; the solution always has the same shape but is stretched out to a greater or lesser extent along the x-axis. Note the way that, *for a given value of x*, the temperature gradually rises to its final unit value.

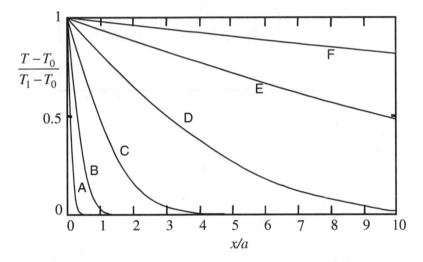

Figure 4.1. The solution for $(T - T_0)/(T_1 - T_0)$, given by Equation (4.3), of the bar raised to a constant temperature at one end as a function of x/a (a is an arbitrary length scale) for $\kappa t/a^2$ with the following values: A: 0.01; B: 0.1; C: 1; D: 10; E: 100; F: 1000.

Solution (4.3) can be used to derive another, equally important solution. Equation (4.1) does not contain x explicitly, so if T is a solution, so is $\partial T/\partial x$. The two facts together mean that

$$\frac{2\kappa}{\lambda\sqrt{4\kappa\pi(t - \tau)}}\exp\left(-\frac{x^2}{4\kappa(t - \tau)}\right)$$

is also a solution for $t > \tau$. Notice that the origin of time has been changed. From this solution it is possible to construct the solution for an arbitrary intensity distribution at $x = 0$. Consider

$$T = T_0 + \int_{\tau=0}^{t} \frac{2I(\tau)\kappa}{\lambda\sqrt{4\kappa\pi(t-\tau)}} \exp\left(-\frac{x^2}{4\kappa(t-\tau)}\right) d\tau. \qquad (4.4)$$

The heat flux is given by

$$F(x,t) = -\lambda \frac{\partial T}{\partial x} = \int_{\tau=0}^{t} \frac{\kappa \lambda I(\tau)}{2\sqrt{\pi}[\kappa(t-\tau)]^{3/2}} \exp\left(-\frac{x^2}{4\kappa(t-\tau)}\right) d\tau.$$

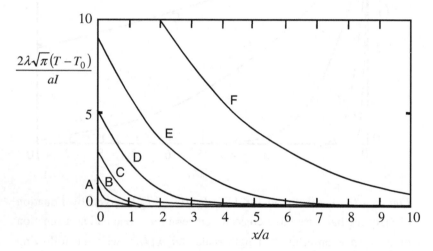

Figure 4.2   The solution   for $2\lambda\sqrt{\pi}\,(T - T_0)/I = \int_{p=0}^{\sqrt{4\kappa t}} \frac{\exp(-x^2/p)}{\sqrt{p}} dp$,

given by Equation (4.5), for a bar with a constant heat flux applied at the end $x = 0$, as a function of x/a (a is an arbitrary length scale) for the same values of $\kappa t/a^2$ as in Figure 4.1; the labels are the same.

Substitute $\tau = t - x^2/4\kappa\lambda^2$ and then take the limit as $x \to 0$ to obtain

$$F(0,t) = -\lambda \frac{\partial T}{\partial x}\bigg|_{x=0} = \frac{2}{\sqrt{\pi}} \int_{\lambda=0}^{\infty} I(t)\exp\left(-\lambda^2\right) d\lambda = I(t).$$

The temperature distribution given by (4.4) is that due to an arbitrarily specified input $I(t)$ with a uniform initial temperature distribution $T = T_0$. In particular, if I is constant, the temperature distribution is given by

$$T = T_0 + \frac{I}{\lambda}\sqrt{\frac{\kappa}{\pi}} \int_{\tau=0}^{t} \frac{1}{\sqrt{\tau}} \exp\left(-\frac{x^2}{4\kappa\tau}\right) d\tau. \tag{4.5}$$

Figure 4.2 shows $2\lambda\sqrt{\pi}\,(T-T_0)/I$ as a function of x for a number of values of $\sqrt{4\kappa t}$. This solution might be used, for example, as the basis of a simple model of the initial stage of the drilling process or in the theory of surface treatment.

4.2 SURFACE HEATING FROM COLD

The time-dependent equation of heat conduction is given by (2.6), and when a thick workpiece in $z \geq 0$ is in steady translation with velocity U in the x direction, it has the form

$$\frac{\partial T}{\partial t} + U\frac{\partial T}{\partial x} = \kappa\left(\frac{\partial^2 T}{\partial x^2} + \frac{\partial^2 T}{\partial y^2} + \frac{\partial^2 T}{\partial z^2}\right). \tag{4.6}$$

Consider the initial value problem that occurs when the laser is switched on at time $t = 0$ and has an incident intensity at the surface given by $I(\mathbf{r},t)$. The initial condition is

$$T(\mathbf{r},0) = T_0$$

and the boundary conditions are

$$-\lambda\frac{\partial T}{\partial z} = (1-\mathcal{R})I(x,y,t) \quad \text{at } z = 0,$$
$$T \to T_0 \quad \text{as } |\mathbf{r}| \to \infty.$$

$\mathcal{R}$ is an appropriate surface reflection coefficient.

Provided an initial value problem such as this is mathematically linear, the use of Laplace transforms[4] can often be a good method of solution. The Laplace transform of a function $f(t)$ is defined by

$$\mathcal{L}(f;\mathbf{r},s) = \int_{t=0}^{\infty} \exp(-st) f(\mathbf{r},t) dt\,,$$

and it has the property (shown by integration by parts) that

$$\mathcal{L}\left(\frac{\partial f}{\partial t};\mathbf{r},s\right) = s\mathcal{L}(f;\mathbf{r},s) - f(\mathbf{r},0).$$

Taking the Laplace transform of Equation (4.6) and making use of the initial condition show that

$$s\mathcal{T} + U\frac{\partial \mathcal{T}}{\partial x} = \kappa\left(\frac{\partial^2 \mathcal{T}}{\partial x^2} + \frac{\partial^2 \mathcal{T}}{\partial y^2} + \frac{\partial^2 \mathcal{T}}{\partial z^2}\right) \qquad (4.7)$$

where $\mathcal{T} = \mathcal{L}(T - T_0;\mathbf{r},s)$. If $\mathcal{J} = \mathcal{L}(I;\mathbf{r},s)$ it satisfies the conditions

$$-\lambda\frac{\partial \mathcal{T}}{\partial z} = (1-\mathcal{R})\mathcal{J} \quad \text{at } z=0, \ \mathcal{T} \to 0 \quad \text{as } |\mathbf{r}| \to \infty.$$

A point source-type solution at the origin can be obtained for Equation (4.7) in exactly the same way as for the time-independent case studied in Section 3.1. Even the argument that provides the multiple corresponding to the power of the source is the same provided the power is replaced by its Laplace transform. It has the form

$$\mathcal{T} = \frac{\mathcal{P}}{2\pi\lambda|\mathbf{r}|}\exp\left(\frac{Ux}{2\kappa} - |\mathbf{r}|\sqrt{\frac{U^2}{4\kappa^2} + \frac{s}{\kappa}}\right) \qquad (4.8)$$

in which $\mathcal{P}$ is the Laplace transform of the power of the point source, $P(t)$, and the form appropriate to a point source at the surface of the

---

[4] Kreyszig, 1993, Ch.6.

workpiece, not in the interior, has been used. The right-hand side without $\mathscr{P}$ is a standard form and is the Laplace transform[5] of

$$\frac{1}{4\pi\lambda|\mathbf{r}|\sqrt{\pi\kappa t^3}}\exp\left[\frac{Ux}{2\kappa}-\frac{U^2t}{4\kappa}-\frac{\mathbf{r}^2}{4\kappa t}\right]. \tag{4.9}$$

Consequently, (4.8) is the product of two Laplace transforms; the inverse of such a product, however, is just the convolution of the original functions[6] so that the inverse transform of (4.8) is

$$\int_{t_1=0}^{t}\frac{P(t-t_1)}{4\pi\lambda|\mathbf{r}|\sqrt{\pi\kappa t_1^3}}\exp\left[\frac{Ux}{2\kappa}-\frac{U^2t_1}{4\kappa}-\frac{|\mathbf{r}|^2}{4\kappa t_1}\right]dt_1. \tag{4.10}$$

The surface convolution method can be used as in Section (3.1) to construct the temperature distribution caused by the given incident intensity $I(x,y,t)$. For this purpose it can be considered a power input of magnitude $(1-\mathscr{R})I(x_1,y_1,t)dx_1dy_1$ at the surface element of magnitude dx_1dy_1 at the point (x_1,y_1,t). The solution to the problem originally posed is

$$T(\mathbf{r},t)=T_0+\int_{-\infty}^{\infty}\int_{-\infty}^{\infty}\int_{t_1=0}^{t}\frac{(1-\mathscr{R})I(x_1,y_1,t-t_1)}{4\pi\lambda|\mathbf{r}-\mathbf{r}_1|\sqrt{\pi\kappa t_1^3}}\times$$

$$\times\exp\left[\frac{U(x-x_1)}{2\kappa}-\frac{U^2t_1}{4\kappa}-\frac{|\mathbf{r}-\mathbf{r}_1|^2}{4\kappa t_1}\right]dt_1dy_1dx_1. \tag{4.11}$$

If time is made dimensionless with respect to $4\kappa/U^2$, all distances with respect to $2\kappa/U$, and $T-T_0$ with respect to $UP/\kappa\lambda$ where P is the total absorbed incident power, Equation (4.9) for a point source at $\mathbf{r}=0$, $t=0$ becomes

$$\frac{2}{(4\pi t')^{3/2}}\exp\left(x'-t'-\frac{\mathbf{r}'^2}{4t'}\right),$$

---

[5] Abramowitz and Stegun, 1965, Formulae 29.2.14 and 29.3.82.
[6] Abramowitz and Stegun, 1965, Formula 29.2.8.

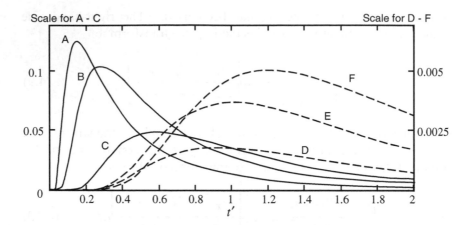

Figure 4.3. Temperature as a function of time for the point source at the origin active only at $t' = 0$, expressed in dimensionless form as a function of time at the following points. Solid line A: (0,0,1); B: (1,0,1); C: (2,0,1); broken lines D: (0,0,3); E: (1,0,3); F: (2,0,3). (The scale for the broken curves is 20 times the scale for the solid curves.)

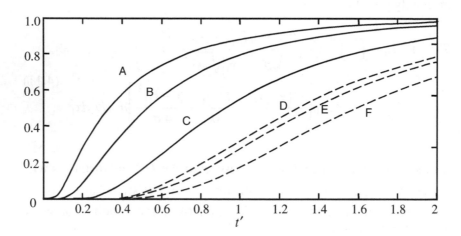

Figure 4.4. Temperature for a constant point source at the origin with transients as a function of time, expressed in dimensionless form as a fraction of the point source asymptotic value, at the same positions as in Figure 4.3. The same scale is used for both sets of curves.

when suitably scaled. Figure 4.3 shows the temperature as a function of time, at a number of different points. It must be remembered that the coordinate system is fixed relative to the point source, not the workpiece. Notice the time lag in the horizontal direction in curves A to C, and in the vertical direction for curves D to F.

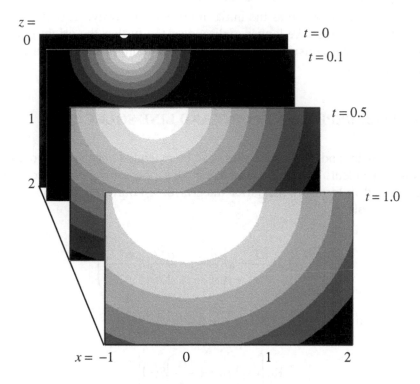

Figure 4.5. Contours of the ratio of the time-dependent solution for a point source at the origin to the asymptotic value. The lighter the shade of the fill, the closer the ratio is to unity.

For a constant point source at the origin suitably scaled, (4.10) gives

$$T' = \int_{t_1=0}^{t'} \frac{2}{(4\pi t_1)^{3/2}} \exp\left[x' - t_1 - \frac{r'^2}{4t_1}\right] dt_1 \, .$$

Figure 4.4 shows examples of the solution given as a fraction of the asymptotic form, which is just the point source solution given as

$$T'_\infty = \int_{t_1=0}^{\infty} \frac{2}{(4\pi t_1)^{3/2}} \exp\left[x' - t_1 - \frac{r'^2}{4t_1}\right] dt_1 = \frac{1}{2\pi|\mathbf{r'}|} \exp(x' - |\mathbf{r'}|).$$

Figure 4.5 shows contours of the same ratio at successive times in the plane of symmetry, $y' = 0$. Notice the way in which the equilibrium solution is approached as the initial transients die away. It will be seen that the transients take longer to disappear the further the origin is from the point under consideration, relative to the point of application of the power source.

4.3 TIME-DEPENDENT POINT AND LINE SOURCES

The point and line source ideas can be extended very simply to cover time-dependent cases by supposing that the power input can be expressed in terms of a Fourier integral or a Fourier series. Look for solutions of

$$\frac{\partial T}{\partial t} + U \frac{\partial T}{\partial x} = \kappa \left(\frac{\partial^2 T}{\partial x^2} + \frac{\partial^2 T}{\partial y^2} + \frac{\partial^2 T}{\partial z^2} \right)$$

of the form

$$\mathrm{Re}\left\{ \exp\left(i\omega t + \frac{Ux}{2\kappa} \right) S(r) \right\}.$$

As before, $r = \sqrt{x^2 + y^2 + z^2}$. Substitution into the equation shows that S must satisfy

$$\left(i\omega + \frac{U^2}{4\kappa} \right) S = \frac{\kappa}{r} \frac{d^2}{dr^2} (rS).$$

The form is identical to that of the equation for S found in the time-independent case, except that the constant on the left has been modified.

Consequently, the appropriate solution that tends to zero as r tends to infinity in all directions is

$$\frac{A}{r}\exp\left\{i\omega t - \frac{Ur}{2\kappa}\sqrt{1+\frac{4\kappa\omega i}{U^2}}\right\}.$$

The square root taken must be the one with a positive real part. Consider the power input at the origin, when the origin is considered to be inside an infinite workpiece in the same way as in Section 3.2, and require this to be the real part of $\exp(i\omega t)$. The multiple A has to be $1/4\pi\lambda$ giving

$$\frac{1}{4\pi\lambda r}\exp\left\{i\omega t + \frac{U}{2\kappa}\left(x - r\sqrt{1+\frac{4\kappa\omega i}{U^2}}\right)\right\}. \tag{4.12}$$

Once again, if the point source is on the surface of a semi-infinite work-piece the expression must be doubled giving

$$\frac{1}{2\pi\lambda r}\exp\left\{i\omega t + \frac{U}{2\kappa}\left(x - r\sqrt{1+\frac{4\kappa\omega i}{U^2}}\right)\right\}. \tag{4.13}$$

In exactly the same way, the corresponding solution for a line source can be obtained. The procedure is as above but, a solution independent of y must be found in terms of $s = \sqrt{x^2 + z^2}$ and x. The only difference from the argument that leads to the solution obtained for the function S given in Section 3.2 is that $U^2/4\kappa$ is replaced by $U^2/4\kappa + i\omega$. The required multiple is found in the same way as for the case that is independent of t and it has the same value. Consequently, the required solution is

$$\frac{1}{2\pi\lambda}\exp\left(i\omega t + \frac{Ux}{2\kappa}\right)K_0\left(\frac{Us}{2\kappa}\sqrt{1+\frac{4\kappa\omega i}{U^2}}\right). \tag{4.14}$$

Solution (4.14) is appropriate for a line source in the interior of a very thick workpiece and, once again, will have to be doubled if it lies in the surface.

Simon's number,

$$Si = \frac{8\pi\kappa}{t_0 U^2},$$ (4.15)

characterizes a periodic solution with period t_0, and appears in the above solutions when ω is identified with $2\pi/t_0$. It was first identified by Simon, Gratzke, and Kroos in their study of a cylindrical keyhole with a periodic incident power.[7]

The functions given by (4.12) and (4.13) are appropriate for one Fourier component of a time-dependent point source, while Equation (4.14) applies in the same way to a line source. It is then possible to construct solutions of time-dependent problems in terms of these point and line source solutions employing all the same techniques that apply to the steady-state case, with the aid of Fourier analysis.

As an illustration of the way in which these solutions can be employed to find the temperature distribution for a time-dependent heat source, consider the problem posed by a heat source at the origin $(0,0,0)$ of Cartesian coordinates. Suppose its absorbed power is periodic with period t_0 and is given in the interval $-\frac{1}{2}t_0 \le t < \frac{1}{2}t_0$ by

$$P(t) = \begin{cases} P_0 t_0/\tau & |t| < \frac{1}{2}\tau \\ 0 & \frac{1}{2}\tau < |t| < \frac{1}{2}t_0 \\ P_0 t_0/2\tau & \text{otherwise.} \end{cases}$$ (4.16)

See Figure 4.6; the mean power of this pulsed input is P_0. Its Fourier series is[8]

---

[7] Simon, Gratzke, and Kroos, 1992. A number of periodic solutions of the equation of heat conduction are solved by Carslaw and Jaeger, 1959, using Fourier series methods (p.105-112) and Laplace transform methods (p.399-401).

[8] Kreyszig, 1993, Ch.10.

$$P(t) = P_0 + \sum_{n=1}^{\infty} a_n \cos \frac{2\pi t}{t_0} = P_0 + \sum_{n=1}^{\infty} a_n \, \text{Re}\left(\exp \frac{2\pi t}{t_0} i \right) \quad (4.17)$$

in which

$$a_n = \frac{2P_0}{\tau} \int_{t=-\tau/2}^{\tau/2} \cos \frac{2\pi t}{t_0} \, dt = \frac{2P_0}{n\pi} \sin \frac{n\pi\tau}{t_0}.$$

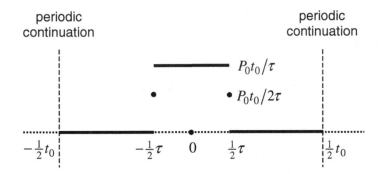

Figure 4.6. The pulsed power input given by (4.16).

It is now possible to use the Fourier series (4.17) for the pulsed power input to find the resulting temperature distribution for this particular point source using Equation (4.13). Thus

$$T(x,r,t) = T_0 + \frac{P_0}{2\pi\lambda r} \exp\left\{ \frac{U}{2\kappa}(x-r) \right\} +$$

$$+ \frac{P_0}{\pi^2 \lambda r} \sum_{n=1}^{\infty} \frac{1}{n} \sin \frac{n\pi\tau}{t_0} \, \text{Re}\left(\exp\left\{ \frac{2n\pi}{t_0} i + \frac{U}{2\kappa}(x - r\sqrt{1 + in\text{Si}}) \right\} \right)$$

$$= T_0 + \frac{P_0}{2\pi\lambda r} \exp\left\{ \frac{U}{2\kappa}(x-r) \right\} +$$

$$+ \frac{P_0}{\pi^2 \lambda r} \sum_{n=1}^{\infty} \frac{1}{n} \sin \frac{n\pi\tau}{t_0} \exp\left\{ \frac{U}{2\kappa}(x - rs_n) \right\} \times$$

$$\times \cos\left(\frac{2n\pi t}{t_0} - \frac{U}{4\kappa} \frac{r \, n\text{Si}}{s_n} \right)$$

$$(4.18)$$

where

$$s_n = \sqrt{\frac{1+\sqrt{1+n^2 Si^2}}{2}} \, .$$

A simpler way to write this is as

$$T = T_0 + \frac{P_0 U}{\lambda \kappa} T'(x', y', z', t')$$

in which

$$x' = Ux/2\kappa, \quad r' = Ur/2\kappa, \quad t' = 2\pi t/t_0$$

and

$$T'(x', y', z', t') = \frac{1}{4\pi r'} \exp(x' - r') +$$

$$+ \sum_{n=1}^{\infty} \frac{2}{n\pi} \sin\left(\frac{mn\pi}{m+1}\right) \frac{1}{4\pi r'} \exp(x' - r's_n) \cos\left(nt' - \frac{r' n Si}{2 s_n}\right)$$

$$(4.19)$$

with $m = \tau : t_0 - \tau$, the ratio of the time during which the laser is on to the time in which it is off. It should be noted that this series converges slowly, so a substantial number of terms may be needed for its evaluation.

Figures 4.7 and 4.8 show isotherms of T' in the y-direction, as t' varies over two periods for the case $Si = 10$ and $m = 1$ just below the surface of the workpiece at $z' = 0.1$ in the case of Figure 4.7. Similarly, isotherms are shown in Figure 4.8 for variation with depth z' at $y' = 0.1$. The effect of altering the value of Si is shown in Figure 4.9. It shows $T'(0,0,0.4,t')$ over one period, $-\pi < t' < \pi$. The value of m is 1, and the values of Si shown are 0, 5, 10, and 20. In the same way, Figure 4.10 shows contours of $T'(0,0,0.4,t')$ as a function of m for Si $= 4$. The values of m shown are $\frac{1}{4}, \frac{1}{2}, 1, 2$.

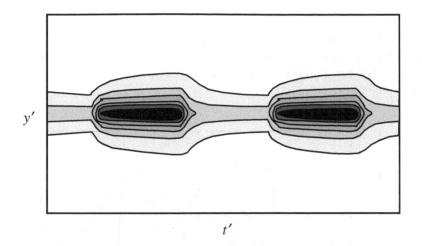

Figure 4.7   Isotherms of $T'(0, y', 0.1, t')$ in the case when $Si = 10$ and $m = 1$ just below the surface, at $z' = 0.1$ with $x' = 0$.

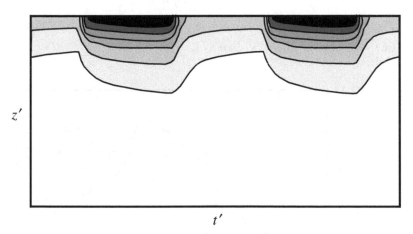

Figure 4.8   Isotherms of $T'(0, 0.1, z', t')$ in the case when $Si = 10$ and $m = 1$ at $y' = 0.1$ with $x' = 0$.

The solution given here for the periodic time-dependent point source can be used as the starting point for the construction of periodic solutions for other incident beam shapes. The method is the same as the one demonstrated in Section 3.2 for the steady case. Equations (3.9) and (3.10) still apply, with the substitution of the periodic solution given by (4.13) in place of the point source solution.   From this, Fourier analysis can be used to obtain the solution when there is a given time dependence within the cycle.   The procedure is no more complicated than it is for the examples worked here, but the resulting expressions tend to be lengthy.

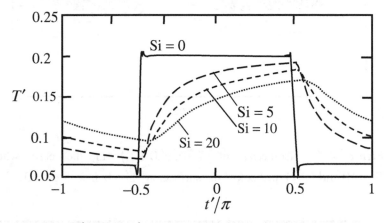

Figure 4.9. $T'(0,0,0.4,t')$ over one period, $-\pi < t' < \pi$ as a function of t' for Si $= 0,5,10,20$ when m is 1.

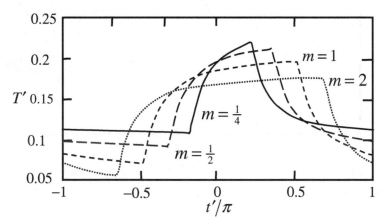

Figure 4.10. $T'(0,0,0.4,t')$ over one period, $-\pi < t' < \pi$ as a function of t' for $m = \frac{1}{4}, \frac{1}{2}, 1, 2$ m when Si is 4.

Equation (4.14) gives the form for the time-dependent periodic line source and it can be used in exactly the same way, but here the new feature is that the Bessel function has a complex argument. Definitions and formulae for the evaluation of Bessel functions with complex arguments can be found in the literature,[9] and the latest versions of some numerical algebra computer packages are capable of evaluating them. They do not lend themselves, however, to solutions that can be written in terms of standard real-valued functions in the way that the point source solution does. There is no simple equivalent, for example, to Equation (4.18) as an alternative to (4.17) as there is for the point source.

4.4 THE THERMAL HISTORY OF A MATERIAL ELEMENT

In a great many problems in material processing it is extremely convenient to use a coordinate system that is fixed with respect to the power source. The reason is that after a time, conditions may become quasi-steady if the power of the source remains constant, as is the case for a laser in a cw mode of working. Dependence on time can then be dropped from the problem once a steady state has been reached, and that leads to a great deal of simplification in the mathematical models. In such a coordinate system, any function of the coordinate vector $\mathbf{r}$ and time t is the value of the function at that particular point and that particular time. This is true irrespective of what particular material element happens to be there at the time, even when conditions have not reached a steady state. Thus, for example, $\mathbf{u}(\mathbf{r},t_1)$ and $\mathbf{u}(\mathbf{r},t_2)$ will in general refer to the velocity of entirely different elements, even though they are measured at the same position in space relative to the heat source. Such a description is usually referred to as an *Eulerian* description and has a great many advantages. It does, however, have the disadvantage that it is hard to track such things as the thermal history of a fluid element.

An alternative way of describing what happens is to use a co-ordinate system that refers everything to the initial position of each material element. Such a description is called a *Lagrangian* description. Suppose that at time $t = 0$ a particular material element is at the point whose coordinates in a fixed Cartesian frame of reference are $\mathbf{r} = \mathbf{a} = (a, b, c)$. Suppose, also, that at a subsequent time t that same

---

[9] Abramowitz and Stegun, 1965. See, for example, Formulae 9.6.13, 9.6.17, 9.6.24, or 9.7.2.

material element is at the position $\mathbf{r} = \xi(\mathbf{a}, t)$. There is then an initial condition $\xi(\mathbf{a}, 0) = \mathbf{a}$; the velocity of this element at time t is then $\dot{\xi}(\mathbf{a}, t)$ where the dot indicates differentiation with respect to time. Suppose the velocity in an Eulerian description of the motion is $\mathbf{u}(\mathbf{r}, t)$ and this is regarded as known (from a solution obtained in the Eulerian frame of reference, for example). It is then possible to find $\xi(\mathbf{a}, t)$ by solving the system of differential equations

$$\frac{\partial}{\partial t}\xi(\mathbf{a}, t) = \mathbf{u}(\xi(\mathbf{a}, t), t), \quad \xi(\mathbf{a}, 0) = \mathbf{a}. \tag{4.20}$$

In general, this is not a particularly simple matter, and normally requires a computed solution rather than an analytical one.

Matters are rather simpler, however, if every element of the work-piece moves with a constant velocity, as will be the case to a good degree of approximation if the workpiece is solid and is moved steadily past the power source.[10] Then

$$\xi(\mathbf{a}, t) = \mathbf{a} + t\mathbf{u},$$

and if the temperature distribution is given in the Eulerian description by $T(\mathbf{r}, t)$, the thermal history of an element initially at $\mathbf{r} = \mathbf{a}$ is simply given by

$$T(\mathbf{a} + t\mathbf{u}, t). \tag{4.21}$$

For example, the temperature at a point one unit below the surface on the center line of a unit point source on the surface of a workpiece traveling at unit velocity is given by

$$T'(t') = \frac{1}{2\pi} \frac{\exp\left([-2 + t'] - \sqrt{[-2 + t']^2 + 1^2}\right)}{\sqrt{[-2 + t']^2 + 1^2}} \tag{4.22}$$

---

[10] The reason for the reservation is that thermal expansion may occur resulting in some small displacements, and it may be of importance to the process being studied.

if it is initially at $x' = -2$ at $t' = 0$; see Equations (3.7) and (4.22). Its graph is shown in Figure 4.11; it is then possible to see such things as the greatest temperature it has reached, the greatest rate of cooling, or the length of time spent in a given temperature range.

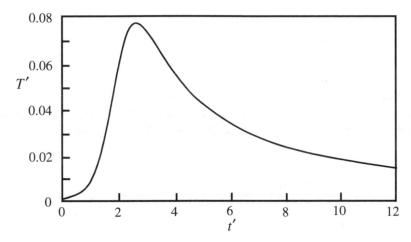

Figure 4.11. Graph of the thermal history of a material element initially at (-2,0,1) whose temperature is given by (4.22).

Use of any standard numerical algebra package shows that the greatest value of T' occurs when $t' = 2.648$; this is after the material element has passed beneath the point source on the surface, and the temperature then is 0.07755 units.   The greatest rate of cooling is 0.01925 units at time 3.411 when the temperature is 0.06694.

Equation (3.7) shows that distances are scaled with $2\kappa/U$ and temperatures with $PU/k\kappa$, so that times are scaled with $2\kappa/U^2$. So if, for example, the power of the laser is 500 W, the speed of translation is $10 \, \text{mm s}^{-1}$ and the material is a stainless steel with a value of $15 \, \text{W m}^{-1} \text{K}^{-1}$ for λ and $2.13 \times 10^{-5} \, \text{m}^2 \text{s}^{-1}$ for κ; these figures mean that at a depth of 4.25 mm the maximum temperature is about 1200 K above the ambient temperature (which would mean that it was molten), and the greatest rate of cooling is $0.37 \times 10^5 \, \text{K s}^{-1}$.

A second example is shown in Figures 4.12 and 4.13. It corresponds to the heated strip whose solution is given in Equations (3.29) and

(3.30). If time is scaled with a/U the thermal history in dimensionless form given by (3.30) at a depth z is

$$f(t', z', \text{Pe}) = \frac{1}{\text{Pe}} \int_{x_1'=t'-\text{Pe}}^{t'+\text{Pe}} \exp x_1' \, K_0\left(\sqrt{x_1'^2 + z'^2} \right) dx_1'. \quad (4.23)$$

The origin of time is chosen so that the particular element is directly under the center of the irradiating beam at the time $t' = 0$. The graphs are exactly the same shape as the graphs of temperature as a function of x', but that is because the motion is uniform. If it is not, as for example when changes in density occur, or when part of the motion is in a fluid state, it will not necessarily follow that this will be the case.

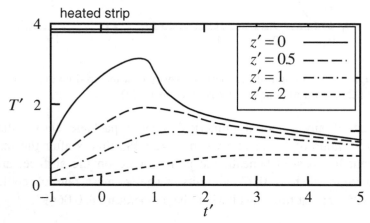

Figure 4.12. Thermal history of material elements at a number of different depths expressed in dimensionless form for Pe = 1. Compare with Figure 3.20.

As an example of thermal history of individual elements in a time-dependent problem, consider the example of Section 4.3 above of a pulsed point source. If the problem is put in dimensionless form, the temperature distribution is given by Equation (4.19) as $T'(x', y', z', t')$. From the scaling factors applied to obtain it from (4.18) it follows that the scale for the velocity of translation U is $\dfrac{2\kappa}{U} \bigg/ \dfrac{t_0}{2\pi} = \dfrac{4\pi\kappa}{Ut_0}$. In dimensionless terms, therefore, the velocity of translation is

$$U \bigg/ \sqrt{\frac{4\pi\kappa}{Ut_0}} = \frac{2}{Si}.$$

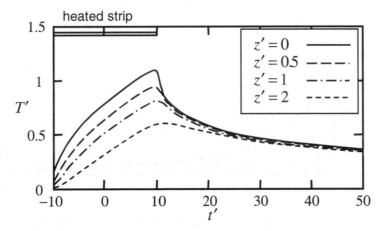

Figure 4.13. Thermal history of material elements at a number of different depths expressed in dimensionless form for Pe = 10. Compare with Figure 3.21.

If a given material element is at $x' = a$, $y' = b$, $z' = c$ at the initial instant $t' = 0$, it will be at $x' = a + 2t'/Si$, $y' = b$, $z' = c$ at a time t'. In dimensionless terms, therefore, its thermal history is given by

$$T'\left(a + \frac{2t'}{Si}, b, c, t'\right). \tag{4.24}$$

Unlike the case of the steady state considered in the previous examples, it is not easy to see what the history is simply by consulting graphs of temperature in the frame of reference that is stationary relative to the power source.   Figure 4.14 shows the temperature distribution $T'(2t'/Si, b, 0, t')$ given by (4.24) for points on the surface for several different values of b, the lateral distance from the plane of symmetry. Time is measured from the instant at which the elements pass the point source, which is at $x' = 0$. The temperature is shown as a graph as a function of elapsed time.   Notice in particular that the maximum temperature moves further to the right as the distance from the plane of symmetry increases.

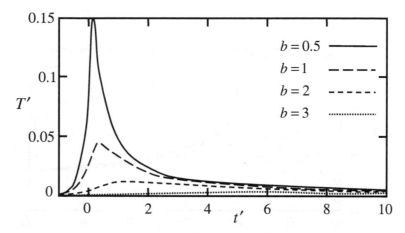

Figure 4.14. Thermal history $T'(2t'/\mathrm{Si}, b, 0, t')$ of material elements in the surface for the pulsed point source. The value of Si is 4, and the value of m is $\frac{1}{9}$.

As an example of the kind of use to which this can be put, expression (4.24) could be used to find the maximum temperature experienced by each material element. This is a relatively difficult operation. For each a,b,c the maximum value of the temperature experienced by that particular element has to be found, either by investigating the zeros of its t-derivative or numerically. It is then possible to draw graphs or contour plots of the maximum temperature experienced by the elements of the workpiece. One could consider cross-sections perpendicular to the laser in figure 4.15, which shows a simple graph of the temperature in the workpiece on the surface. With this particular example, the maximum temperature is infinite but the graph has been truncated. Such information could be used to see which parts of the material reach melting temperature or some other critical phase transition. Other quantities, such as the maximum rate of change of temperature, or more complicated criteria can be investigated in the same way. Numerical examples such as this show how useful such methods can be.

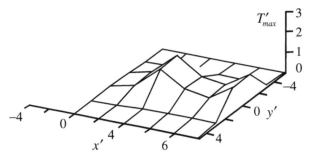

Figure 4.15. A simple illustration of the results of an investigation of the way in which the maximum temperature experienced by any element of the workpiece may be calculated. The maximum temperature at the surface is shown for the example of (4.24). The input power is pulsed and the graph is spatially periodic. With a suitable value for the melting isotherm (with the parameters used in the given example, the melting isotherm corresponds to a value if 1.67 for steel) it is possible to construct this isotherm.

CHAPTER 5

MOVING BOUNDARY PROBLEMS

5.1 STEFAN PROBLEMS

There are special effects at any change in phase of a material that require the input or release of heat. Familiar examples are the additional energy needed to melt water, or the scalding caused by the condensation of steam as it releases its latent heat. The same may be true at less obvious transitions between different metallurgical states.[1] From the point of view of thermal modeling these are all very similar, involving as they do the solution of a problem in heat conduction with an internal boundary (drilling problems are similar,[2] as is laser cutting.[3]) The location of the boundary is not known in advance, and is, in fact, a part of the solution of the problem. Problems of this kind are often referred to as *Stefan problems*.

As an illustration, consider a long uniform column of cross-sectional area A thermally insulated from its surroundings. The top is filled with water whose surface is at $z = Z_B(t)$. Its surface is being boiled away by an incident intensity I kW s^{-1}. It will be assumed constant and that all of it is absorbed. Below is a layer of water, and below that again a layer of ice whose temperature very far down has the constant value $T_0 < T_M < T_B$, where T_M is the melting temperature of water and T_B its boiling temperature. The melting boundary is assumed to be at $z = Z_M(t)$. Since the ice is cooler than the water it can conduct heat away from the interface between the two and, consequently, there is a moving boundary of melting water advancing into the ice. See Figure 5.1 for a diagram of the geometry of the problem.

---

[1] Metiu et al., 1976 and 1987.
[2] Andrews and Atthey, 1975 and 1976; Solana et al., 1999.
[3] Vicanek and Simon, 1987; Vicanek et al., 1987.

In general, this is a relatively complicated initial value problem, but if time is allowed to elapse sufficiently, a quasi-steady state can occur in which both the boiling and melting boundaries move downward with a constant velocity U. Suppose that the surface is at $z = 0$ at $t = 0$ and the melting boundary at $z = a$. Subsequently, they are at $z = Z_B(t) = Ut$ and $z = Z_M(t) = a + Ut$, respectively.

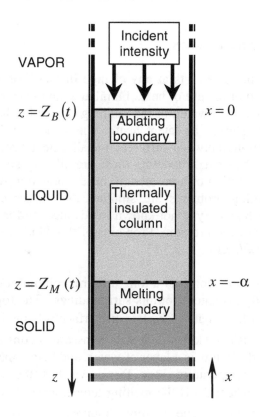

Figure 5.1. The moving boundary between ice and water.

The temperature in the frozen region satisfies the one-dimensional equation of heat conduction[4]

$$\frac{\partial T}{\partial t} = \kappa \frac{\partial^2 T}{\partial z^2},$$

(5.1)

---

[4] From Equation (2.6)

the two surface boundary conditions

$$T = T_B \qquad \text{on} \quad z = Ut,$$

$$-\lambda \frac{\partial T}{\partial z} + \rho L_V U = I \quad \text{on} \quad z = Ut,$$

two conditions on the melting interface

$$T = T_M \qquad \qquad \text{on} \quad z = a + Ut,$$

$$\left[-\lambda \frac{\partial T}{\partial z} \right]_{ice}^{water} = \rho L_M U \quad \text{on} \quad z = a + Ut,$$

and the asymptotic condition

$$T \to T_0 \quad \text{as} \quad z \to \infty.$$

The second of each pair of conditions express the requirement that the latent heat L_V released on boiling and L_M released on freezing are conducted away entirely into the ice. Notice that there are two more conditions than one would normally expect for a second-order equation. The additional conditions are necessary since neither the location of the boiling surface nor the melting interface is known in advance.

The easiest way to solve the problem is to look for a temperature distribution of the form

$$T = T_M + (T_V - T_M)f(x) \text{ in the water}$$
$$T = T_0 + (T_M - T_0)g(x) \text{ in the ice}$$

where

$$x = Ut - z.$$

For simplicity, take the thermal constants of ice and water to be equal.[5] Then f and g have to satisfy

$$U \frac{df}{dx} = \kappa \frac{d^2 f}{dx^2},$$

---

[5] But compare this approximation with the values quoted in Appendix 1.

$$U\frac{dg}{dx} = \kappa\frac{d^2g}{dx^2},$$
$$f(0)=1,$$
$$f(-a)=0,$$
$$g(-a)=1,$$
$$g\to0 \quad \text{as} \quad x\to-\infty$$

in order to satisfy the equation of heat conduction and the temperature boundary conditions. The two ordinary differential equations have general solutions which are the sum of one arbitrary constant added to a second arbitrary constant multiplying $\exp(Ux/\kappa)$. The four boundary conditions given above for f and g show that

$$f(x) = \frac{\exp[U(x+a)/\kappa]-1}{\exp(Ua/\kappa)-1}$$

and

$$g(x) = \exp[U(x+a)/\kappa].$$

The two conditions on the heat flux require, respectively, that

$$f'(0) = \frac{I-\rho L_V U}{\lambda(T_V - T_M)}$$

and

$$(T_V - T_M)f'(-a) - (T_M - T_0)g'(-a) = \frac{\rho L_M U}{\lambda}.$$

The first of these together with the solution for f shows that

$$U\left\{\rho L_V + \frac{\lambda(T_V - T_M)}{\kappa[1-\exp(-Ua/\kappa)]}\right\} = I, \qquad (5.2)$$

while the second shows that

$$\exp\left(\frac{Ua}{\kappa}\right) = 1 + \frac{T_V - T_M}{T_M - T_0 + \rho\kappa L_M / \lambda}.$$ (5.3)

Conditions (5.2) and (5.3) together define U and a in terms of I, T_0 and the material constants of ice and water. Their structures are easier to see if they are made dimensionless. Put

$$a' = \frac{Ua}{\kappa}, \ U' = \frac{U}{\sqrt{L_V}}, \ T_0' = \frac{T_M - T_0}{\lambda} \ \text{and} \ I' = \frac{I}{\rho L_M \sqrt{L_V}},$$

then (5.3) shows that

$$a' = \ln\left(\frac{1 + c + T_0'}{1 + T_0'}\right) \text{ where } c = \frac{T_V - T_M}{\rho\kappa L_M}\lambda.$$ (5.4)

Similarly, (5.2) combined with (5.3) to eliminate a' shows that

$$U' = \frac{I'}{1 + c + d + T_0'} \text{ where } d = \frac{L_V}{L_M}.$$ (5.5)

Hence (5.4) and (5.5) show that there are quite simple relations between the thickness of the liquid region, the rate of ablation, and the temperature at depth of the frozen region and the incident intensity.

With values appropriate to water $c = 1.27$ and $d = 6.82$. With these values, Figure 5.2 shows the forms of a' and U'/I' as functions of T_0'. As a specific illustration, Figure 5.3 shows the graph of the temperature distribution in dimensional form in the case when T_0 is $-18°C$, i.e., approximately $0°$ on the Fahrenheit scale. The constants chosen approximate to the average for those for water and ice. With an incident intensity of 1 W cm^{-2}, the rate of advance is $U = 5 \text{ mm hr}^{-1}$ and the distance between the melting interface and the boiling boundary is 4 cm.

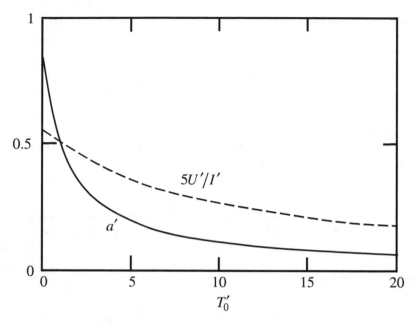

Figure 5.2. a' and U'/I' as functions of T_0'.

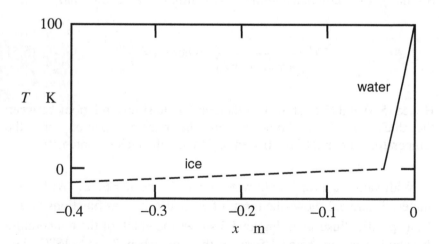

Figure 5.3. The temperature distribution in the moving boundary problem between ice and water whose surface is being boiled away, with $T_0 = -18°C$ and an incident intensity of 1 W cm^{-2}.

Although the example has been presented in terms of water and ice, the same idea is directly applicable to a simple model of laser drilling in which lateral conduction is ignored. It also illustrates the way in which the location of the boundary between different phases is a part of the solution of Stefan problems. This is also entirely characteristic of welding problems. A further point to notice is that the latent heat has a central part to play. By way of illustration, we can consider the same problem but ignore the latent heat of melting. This is equivalent to setting it equal to zero in the above analysis. The solution for the temperature is then given by

$$T = T_0 + (T_V - T_0)\exp\left(\frac{Ix}{\lambda + \rho\kappa L_V}\right).$$

It will be seen that there are differences from the solution in which latent heat is included. It is quite common in simple analytical models of welding processes or thermal treatment, for example, to ignore the contribution due to the latent heat of melting or other phase changes. There is often some justification for the approximation, but it is necessary to be aware that it *is* an approximation and its accuracy should be verified in the given context.

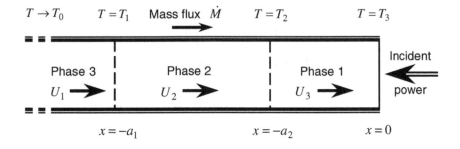

Figure 5.4. The three-phase problem in the steady state.

The same technique can be applied once a steady state has been reached when the material constants of the different phases are not the same, or when there are more than two phases. The problem becomes more complex notationally, but the general principles are the same. Thus, in a frame of reference in which the ablating boundary is at

$x = -a_3 = 0$ and there are three phases, distinguished by subscripts 3, 2, and 1 counting back from the ablating boundary as shown in Figure 5.4, with interfaces at $x = -a_1$ and $x = -a_2$. Suppose that the temperature of the phase transitions are T_1, T_2, and T_3 with corresponding latent heats L_1, L_2, and L_3 where $T_0 < T_1 < T_2 < T_3$. The differential equations satisfied are

$$U_1 \frac{\partial T}{\partial x} = \kappa_1 \frac{\partial^2 T}{\partial x^2} \quad \text{in } x < -a_1,$$

$$U_2 \frac{\partial T}{\partial x} = \kappa_2 \frac{\partial^2 T}{\partial x^2} \quad \text{in } -a_1 < x < -a_2,$$

and

$$U_3 \frac{\partial T}{\partial x} = \kappa_3 \frac{\partial^2 T}{\partial x^2} \quad \text{in } -a_2 < x < 0.$$

There is now an additional pair of interface conditions representing conservation of the mass flux; all these do is relate the velocities in the three regions. They can be written as

$$\rho_1 U_1 = \rho_2 U_2 = \rho_3 U_3 = \dot{M}.$$

The boundary conditions at $x = 0$ are now

$$T = T_3 \text{ and } \lambda_3 \frac{\partial T}{\partial x} + \rho_3 L_3 U_3 = I \quad \text{at} \quad x = 0,$$

the two pairs of interface conditions are

$$T = T_1 \quad \text{with } T \text{ continuous and } \lambda_1 \frac{\partial T}{\partial x}\bigg|_1 + \dot{M} L_1 = \lambda_2 \frac{\partial T}{\partial x}\bigg|_2 \quad \text{on} \quad x = -a_1,$$

$$T = T_2 \quad \text{with } T \text{ continuous and } \lambda_2 \frac{\partial T}{\partial x}\bigg|_2 + \dot{M} L_2 = \lambda_3 \frac{\partial T}{\partial x}\bigg|_3 \quad \text{on} \quad x = -a_2,$$

and

$$T \to T_0 \quad \text{as} \quad x \to -\infty.$$

The solutions of the three partial differential equations are all of the form

$$T = A_i + B_i \exp(U_i x / \kappa_i) \quad \text{for} \quad i = 1..3.$$

The boundary and interface conditions determine the six constants A_i and B_i, the location of the interfaces a_i, and the mass flow rate $\dot{M}$.

Some alloys and rocks do not have definite values for their melting points but melt over a temperature range. Suppose this range is $T_1 < T < T_2$. There are a number of ways in which the problem this poses can be tackled, depending on the degree of sophistication required, but a simple approach is as follows.[6] Suppose the latent heat required for melting is L_1 and the specific heat in this range is c_2'. Then the latent heat could be allowed for by replacing c_2' with $c_2 = c_2' + L/(T_2 - T_1)$, which is equivalent to the assumption that the latent heat is absorbed uniformly over the melting range. In that case the problem is reduced to one in which there is no latent heat at the two internal boundaries, but here is a modified specific heat. The problem is therefore exactly as above, but with L_1 and L_2 set zero and the thermal diffusivity $\lambda_2 / \rho_2 c_2'$ replaced by

$$\kappa_2 = \lambda_2 / \rho_2 [c_2' + L/(T_2 - T_1)].$$

An example of a problem such as this is the melting of a solder wire by a hot soldering iron, although the boundary conditions are somewhat different. The end to which the iron is applied is at a fixed temperature that is higher than the temperature of the wire. The melting zone then moves gradually along the length of the wire. It is not to be expected that its velocity is independent of time, and a somewhat different technique is necessary for its solution.

The first solution to be found for problems of change of state was of this type. It is known as Neumann's problem.

---

[6] Carslaw and Jaeger, 1959, p.289.

5.2 NEUMANN'S PROBLEM[7]

Neumann's solution exploits the properties of the time-dependent solution (4.3) of the equation of heat conduction in one dimension, Equation (5.1), which is a special case of Equation (2.6). The error-function solution has special properties that make it very useful for problems of this type.

Suppose that there is a liquid in $x > 0$ whose temperature is initially $T_1 > T_0$, where T_0 is the temperature to which the boundary $x = 0$ is reduced at time $t = 0$, and T_F is the freezing temperature of the liquid. In the solid phase the material has properties distinguished by the subscript S, and in the liquid phase by the subscript L. The density ρ will be assumed to be the same in both phases. The material is at rest and the boundary is moving. Its position will be taken to be at $x = X(t)$ at time t. The boundary condition at $x = 0$ is

$$T(0,t) = T_0 ;$$

the conditions at the freezing boundary are

$$T_{solid}(X(t),t) = T_{liquid}(X(t),t) = T_F ,$$

$$\lambda_S \left.\frac{\partial T}{\partial x}\right|_{solid} - k_L \left.\frac{\partial T}{\partial x}\right|_{liquid} = \rho L \frac{dX}{dt} \text{ at } x = X(t),$$

and the asymptotic condition is

$$T_{liquid} \rightarrow T_1 \text{ as } t \rightarrow \infty$$

with

$$X(0) = 0, \quad T_{liquid}(x,0) = 0 .$$

---

[7] Given by F. Neumann in his lectures in the 1860's; Weber and Riemann, 1919, p.121. Part of the solution of the same problem was published by Stefan, 1891.

The technique for solution is to use the properties of solutions like (4.3) and seek forms for the temperature distribution such as

$$T_{solid} = T_0 + A \operatorname{erf} \frac{x}{\sqrt{4\kappa_S t}}$$

and

$$T_{liquid} = T_1 - B \operatorname{erfc} \frac{x}{\sqrt{4\kappa_L t}} .$$

The first of these satisfies the condition at $x = 0$, and the second satisfies the condition at $x \to \infty$. The temperature interface condition at $x = X$ shows that if there is to be a solution of the type that has been guessed, then, necessarily, X must be proportional to $\sqrt{t}$. Write

$$X(t) = \xi \sqrt{4t \sqrt{\kappa_L \kappa_S}} \; ; \tag{5.6}$$

in that case A and B are given in terms of ξ by

$$A = \frac{T_F - T_0}{\operatorname{erf}\left[\xi(\kappa_L/\kappa_S)^{\frac{1}{4}}\right]} \quad \text{and} \quad B = \frac{T_1 - T_F}{\operatorname{erfc}\left[\xi(\kappa_S/\kappa_L)^{\frac{1}{4}}\right]} .$$

The heat flux interface condition requires that

$$\frac{A\lambda_S \exp\left(-X^2/4\kappa_S t\right)}{\sqrt{\pi\kappa_S t}} - \frac{B\lambda_L \exp\left(-X^2/4\kappa_L t\right)}{\sqrt{\pi\kappa_L t}} = \rho L \frac{dX}{dt} ,$$

and substitution of the values found for A and B and of the form anticipated for X shows that there is indeed a solution of the problem of this type provided that ξ is a solution of the equation

$$\frac{(T_F - T_0)c_{pS}}{L\sqrt{\pi}}\left(\frac{\kappa_S}{\kappa_L}\right)^{\frac{1}{4}}\frac{\exp\left(-\xi^2\sqrt{\kappa_L/\kappa_S}\right)}{\mathrm{erf}\left[\xi(\kappa_L/\kappa_S)^{\frac{1}{4}}\right]} -$$

$$-\frac{(T_1 - T_F)c_{pL}}{L\sqrt{\pi}}\left(\frac{\kappa_L}{\kappa_S}\right)^{\frac{1}{4}}\frac{\exp\left(-\xi^2\sqrt{\kappa_S/\kappa_L}\right)}{\mathrm{erfc}\left[\xi(\kappa_S/\kappa_L)^{\frac{1}{4}}\right]} = \xi.$$

$$(5.7)$$

The temperature distribution is then given by

$$T_{solid} = T_0 + \frac{T_F - T_0}{\mathrm{erf}\left[\xi(\kappa_L/\kappa_S)^{\frac{1}{4}}\right]}\mathrm{erf}\frac{x}{\sqrt{4\kappa_S t}}$$

in

$$0 < x < X(t) = \xi\sqrt{4t\sqrt{\kappa_L \kappa_S}} \qquad (5.8)$$

and

$$T_{liquid} = T_1 - \frac{T_1 - T_F}{\mathrm{erfc}\left[\xi(\kappa_S/\kappa_L)^{\frac{1}{4}}\right]}\mathrm{erfc}\frac{x}{\sqrt{4\kappa_L t}} \quad \text{in } X(t) = \xi\sqrt{4t\sqrt{\kappa_L \kappa_S}} < x.$$

$$(5.9)$$

Figure 5.5 shows the solution given by Equations (5.6) to (5.9) for the case of water when T_0 is $-10°C$ and T_1 is $2°C$ for a number of different values of the time. Notice that the freezing front moves with a steadily decreasing speed to an asymptotic value of zero. Ultimately, however, liquid at any given distance from the origin will freeze, although it may take a very long time to do so. Figure 5.6 shows the temperature at distances of 1 cm, 2 cm, and 3 cm from the origin as a function of time. If meters and seconds are used for x and t, the value of ξ given by Equation (5.7) is 0.2209, and T is given in degrees Centigrade by

$$T = \begin{cases} -10 + 67.65\,\mathrm{erf}\dfrac{470.4x}{\sqrt{t}} & \text{for } x < 2.801\times10^{-4}\sqrt{t} \\[2mm] 2 - 3.331\,\mathrm{erfc}\dfrac{1322x}{\sqrt{t}} & \text{otherwise.} \end{cases}$$

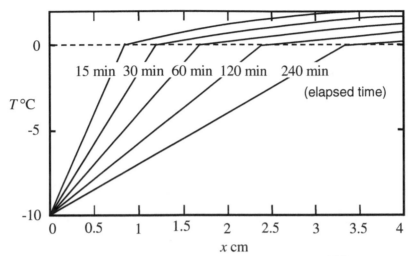

Figure 5.5. Solution for the freezing of water initially at 2°C with one boundary held at −10°C as a function of *x*, at times of 15, 30, 60, 120, and 240 min.

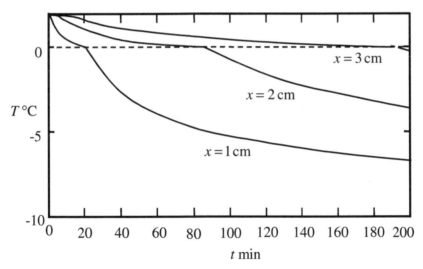

Figure 5.6 The temperature at a distance from the origin of 1 cm, 2 cm, and 3 cm for the same case as that shown in Figure 5.5.

A number of variants of the same problem can be solved in the same way.[8]

---

[8] Carslaw and Jaeger, 1959, 282-294; see for comparison Malmuth, 1976.

CHAPTER 6

SIMPLE MODELS OF LASER KEYHOLE WELDING

_____

6.1 LASER KEYHOLE WELDING

Some of the techniques used in the simpler mathematical models that have been found to be of use in laser welding will be introduced in this chapter. In particular, it will be shown how the point and line source solutions introduced in Chapter 2 and associated with the name of Rosenthal can be used to obtain simple descriptions of the temperature in a workpiece. From these, more elaborate models can be developed to obtain insight before more complicated models need to be constructed. We shall then look at some more advanced but useful analytical models of the solid and liquid phases, and consider features of the dynamics of the keyhole.

In laser welding, as in surface heat treatment and laser cutting, for example, coherent light from the laser forms a spot that can be a concentrated source of heat or perhaps a rather more diffuse region of heating. It is often helpful in some of these applications to model the action of the laser as a point source on the surface of the workpiece. In the case of keyhole welding, however, a line source solution and modifications of that idea can be useful for simulating the penetration of the laser into the workpiece.

The line source simulates a laser-generated keyhole. It can be either partially or fully penetrating. It can be of variable length in practice and either constant or varying in strength with depth, but in that case a modification to the basic theory is needed. More complicated sources can be constructed. The standard models treat the case of a workpiece that translates with a uniform velocity U with respect to the laser, as in laser welding.

Energy absorption by the workpiece from the laser can involve a direct interaction with the laser light incident on a surface. This process is usually referred to as Fresnel absorption (see Section 3.4); the absorbed laser light can, however, vaporize material to form a keyhole that may contain ionized vapor. Indirect absorption then occurs by the following two-stage process. Primary absorption takes place in the vapor in the keyhole, which is then transmitted to the walls of the keyhole by thermal conduction in the vapor and other processes.

The laser energy enters the workpiece from the walls of the keyhole, which to a first approximation can be thought of as a circular cylinder, and melts the material by the process of heat conduction forming a weld pool. Complicated motion of the material usually occurs in this weld pool. The three regions are shown schematically in Figure 6.1. The boundary of the molten region of the weld pool is identified by the melting isotherm of the material. Beyond this lies the solid material. Heat can be transported in the liquid molten region by both convection and conduction processes; in the solid region heat is transported relative to the material of the workpiece by conduction processes only.

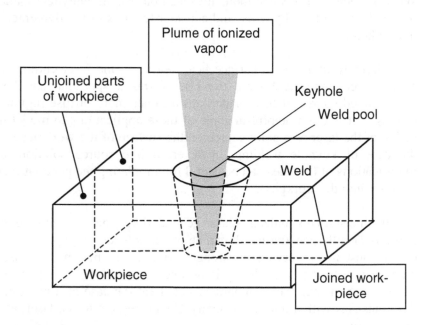

Figure 6.1. Schematic diagram of the solid, liquid, and vapor regions in laser keyhole welding.

A useful parameter for describing conditions is the Péclet number Pe defined by

$$\text{Pe} = \frac{\rho U \ell c_p}{\lambda} \equiv \frac{U\ell}{\kappa} \text{ with } \kappa = \frac{\lambda}{\rho c_p}.$$

Here ρ is the density, U is the velocity of translation, c_p is the specific heat of the material of the workpiece at constant pressure, and λ is the thermal conductivity of the workpiece; κ is the thermal diffusivity. In this expression ℓ is a characteristic length scale of the process. It might be the beam radius of the laser, for example, or one of the principal dimensions of the weld pool. The relevant scale is likely to depend on the particular aspect of interest in the process being studied. Low translation speeds of the workpiece are usually associated with small values of the Péclet number. It is a dimensionless measure of the relative importance of convected to conducted heat.

The Prandtl number Pr is defined by

$$\text{Pr} = \frac{\mu c_p}{\lambda} \equiv \frac{\nu}{\kappa} \text{ with } \nu = \frac{\mu}{\rho}$$

where μ is the (dynamic) viscosity of the molten material and ν is known as its kinematic viscosity. The viscosity is a measure of the diffusion of momentum, and the thermal conductivity is a measure of the diffusion of heat, so the Prandtl number is of significance in problems of convective heat transfer. The equation of heat conduction is central, and from Equation (2.6) is

$$\rho c_p \left(\frac{\partial T}{\partial t} + u \frac{\partial T}{\partial x} + v \frac{\partial T}{\partial y} + w \frac{\partial T}{\partial z} \right) = \lambda \left(\frac{\partial^2 T}{\partial x^2} + \frac{\partial^2 T}{\partial y^2} + \frac{\partial^2 T}{\partial z^2} \right). \quad (6.1)$$

T is the temperature of the material of the workpiece, t is the time, and (u, v, w) is the velocity of its translation relative to the laser beam. It is customary, though not essential, to use a frame of reference in which the laser beam is stationary. The full set of equations governing the motion of the solid and liquid phases of the workpiece, and the transfer of thermal energy within it, are given by Equations (2.6) and (2.24). A number of simplifying assumptions are commonly made, however. One of these is that all material parameters are constant. In reality, such things as the density, the conductivity, etc. will depend on temperature T, but inclusion of this feature is often not justified by the degree of accuracy of information required or available.

An additional approximation is to assume that in a frame of reference stationary with respect to the laser, a steady state has been achieved.

Yet another approximation is to ignore the complicated fluid mechanical motion in the weld pool and thermally induced changes of volume in the solid phase of the workpiece, and assume that all the material is moving unidirectionally with the same velocity U.

A further set of assumptions often made is to suppose that the latent heat of fusion does not play a major role in determining such things as the weld width, and that the parameters of the material have the same values in the liquid and solid states.

These are very substantial assumptions, but nonetheless their use can lead to helpful insights and results that are not necessarily quantitatively very wrong if carefully used. The equation of heat conduction under assumptions of this kind then takes the relatively simple form

$$U \frac{\partial T}{\partial x} = \kappa \nabla^2 T \equiv \kappa \left(\frac{\partial^2 T}{\partial x^2} + \frac{\partial^2 T}{\partial y^2} + \frac{\partial^2 T}{\partial z^2} \right) \qquad (6.2)$$

where κ is the thermal diffusivity and the positive x-axis is in the direction of motion of the workpiece.

6.2 POINT AND LINE SOURCE MODELS

The point source and line source solutions of Rosenthal[1] introduced in Chapter 3, Equations (3.5) and (3.22), are two very simple exact solutions of the equation of heat conduction. They give the temperature field in the workpiece when the source of power is in steady relative motion and a steady state has been achieved. From the time of their discovery, they have been very influential as simple models that give the temperature and the power absorption of a uniform, infinite, or semi-

---

[1] Rosenthal, 1941; Rosenthal, 1946.

infinite workpiece. The point source can be used as a model of a simple conduction weld, and the line source as a model of keyhole welding. There are, however, limitations to them. A particularly obvious example in the context of the use of lasers for welding is the failure of the line source to describe the "nail-head" shape of the cross-section of the fusion zone in the case of the keyhole weld, as illustrated by the examples shown in Figure 6.2. A line source model by itself could lead only to a parallel-sided weld of infinite extent, while a point source model would lead to a weld that is approximately semicircular in cross section. The weld in Figure 6.2(a) has a distinct nail-head section at the top, a feature that is clearly present, though to a much less extent, in the weld of Figure 6.2(b). In neither weld is the continuation downward parallel-sided.

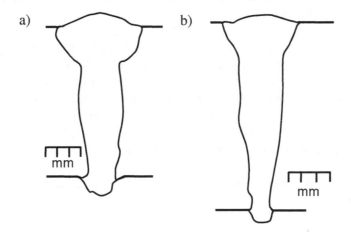

Figure 6.2. Cross-sections of two completed demonstration welds. (a) Weld in stainless steel at 0.5 cm s^{-1} at a power of 9.9 kW. (b) Weld in 1.25 cm-thick C-Mn microalloyed low-oxygen steel plate with a laser power of 6 kW and a speed of 0.58 cm s^{-1}.

The insight that has allowed the development of simple ideas such as this into more realistic tools for the description of welding was the suggestion that a superposition of a point and a line source might provide a more satisfactory description of a keyhole weld. The idea proved to be extremely successful and has led to numerous adaptations and elaborations. For example, multiple point sources can be included to represent absorption occurring preferentially in other parts of the work-piece than just at the surface. A nonuniform but continuous distribution of sources can be used to model the variation of width of the fused material of the weld, and the same idea can be extended to allow for curvature in the keyhole.[2] Similarly, they can be used to describe a nonuniform input over a region of the surface. It is even possible to extend the idea at any of these levels of sophistication to describe time-dependent phenomena, and pulsed welding in particular. The virtue of these models is their simplicity compared to the full formulation of such problems in theoretical form, which normally requires the use of sophisticated software. It is possible to gain valuable qualitative insight by these methods. Such ideas can be used to describe the power absorption characteristics of a given weld as a function of depth or, alternatively, with the aid of simple models of the absorption process, to predict the cross-section of the weld. The temperature field obtained from them can be used to identify the heat-affected zone, or to estimate metallurgical properties.

In the point and line source models of laser keyhole welding, there can be a basic uniform line source with the addition of further point sources. There could be one on the surface whose effect is given by Equation (3.7), one or more in the interior of the workpiece, using formula (3.2), or even a point source above the surface of the work-piece, a modification that has been used, for example, to describe the modeling of the laser welding of copper.[3] See Figure 6.3.

The temperature distribution that results from the configuration shown is

$$T = T_0 + \frac{Q}{\lambda} \text{line}\left(\frac{Ux}{2\kappa}, \frac{Uy}{2\kappa}\right) + \frac{U}{2\kappa\lambda}\left\{ 2P_1 \text{ point}\left(\frac{Ux}{2\kappa}, \frac{Uy}{2\kappa}, \frac{Uz}{2\kappa}\right) + \right.$$

$$\left. + P_2 \text{ point}\left(\frac{Ux}{2\kappa}, \frac{Uy}{2\kappa}, \frac{U(z - z_0)}{2\kappa}\right) + P_2 \text{ point}\left(\frac{Ux}{2\kappa}, \frac{Uy}{2\kappa}, \frac{U(z + z_0)}{2\kappa}\right)\right\}.$$

$$(6.3)$$

The functions point and line are defined in (3.6) and (3.23), respectively. The definitions of P_1, P_2, and Q in the definitions of the strengths of the sources require them to be the power that is absorbed

---

[2] Kaplan, 1997.
[3] Gouveia et al., 1995.

into the workpiece. Any power reflected must be separately accounted for.

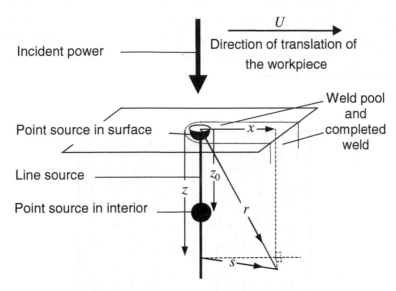

Figure 6.3. Schematic representation of the relative configuration of power source, point source, and line source with the coordinate system employed here.

Notice that in order to allow for a surface at which no heat is lost, so that

$$\frac{\partial T}{\partial z} = 0 \text{ at } z = 0,$$

it has been necessary to introduce the fourth term on the right of (6.3), which is in fact the *image* of the third term. Note that the axis of the line source is perpendicular to the plate.

Figure 6.4 shows isotherms of this in the case when $P_1 = 400$ W, $P_2 = 0$, and $Q_1 = 2.3$ kW cm^{-1} with $U = 5.8$ mm s^{-1}, so that the formula represents a keyhole weld with the "nailhead" feature. The section shown is about 0.3 mm behind the axis of the laser and corresponds approximately to the widest part of the molten region. The parameters are roughly those appropriate to the example of Figure (6.2b). The values for the thermal conductivity and diffusivity used were those for the liquid rather than the solid phase. It will be seen that the total power absorbed into the workpiece, about $3\frac{1}{4}$ W, is a good deal less than the nominal laser power. This figure does not include the power needed to ablate material from the keyhole wall or power lost from the ends of the keyhole. There is a general qualitative similarity on the assumption of an absorption coefficient of the order of 0.5. The nail-head feature is not represented too well however, and the parallel-sided bottom part of the keyhole is very unrealistic.

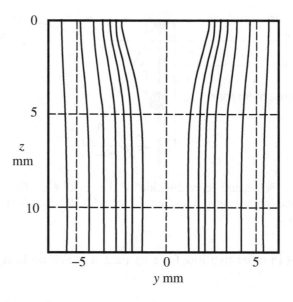

Figure 6.4. Example of a profile constructed using a single point source the surface, and a line source. Isotherms are at 100 K intervals from 1100 K to 1700 K.

The original application was to welds in thick specimens with a point source at the surface and a uniform line source. For this purpose the condition of zero heat loss from the bottom of the workpiece was unimportant. As a qualitative description it was excellent, but there were quantitative discrepancies, the most notable of which was the fact that the lower part of the weld is not parallel-sided. The addition of a point source in the interior can be useful in an attempt to describe the welding of zinc-coated steel sheets.[4] For this purpose a second point

---

[4] Akhter et al., 1989

source in the interior of the weld is a natural extension to the original idea to represent the absorption and ablation of the zinc coatings between the two layers. The problems posed by both features can be overcome in the same way. The constant line source is clearly inappropriate but can be replaced by a continuous distribution of point sources whose strengths vary with position along the line, giving rise to a line source of varying strength. A problem, however, is that the boundary conditions at the surface and bottom of the workpiece are not satisfied. On the assumption that no heat is lost through these surfaces, the difficulty is easily overcome by employing image sources. For each point source and each of its images there must be a reflection in each boundary, leading to a complete infinite image system that can be represented schematically as in Figure 6.5. All the images are of the same strength in the heat theory examples considered here, in which the boundaries are parallel planes.

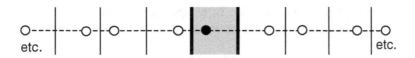

Figure 6.5. The image system for an interior point source in a plane-sided metal sheet, here oriented vertically. The black spot represents the real point source, the shaded region the workpiece, the hollow circles the image sources and the thin lines the image boundaries of the workpiece.

It will be noticed that a point source at the surface can be regarded as the limiting case when a point source just below the surface and its image just above the surface are forced to coincide. This is another way of understanding the origin of the factor 2 in Equation (3.7). If the workpiece lies in $0 \le z \le h$ and the point source is at $z = z_0$ where $0 \le z_0 \le h$, the solution for the temperature in the plate is then given by

$$T =$$

$$T_0 + \frac{PU}{2\kappa\lambda} \sum_{n=-\infty}^{\infty} \{\mathrm{point}(x', y', z' - z_0' + 2nh') + \mathrm{point}(x', y', z' + z_0' + 2nh')\}$$

where

$$x' = \frac{Ux}{2\kappa}, \; y' = \frac{Uy}{2\kappa}, \; z' = \frac{Uz}{2\kappa}, \; z'_0 = \frac{Uz_0}{2\kappa} \text{ and } h' = \frac{Uh}{2\kappa}.$$

It will be noticed that the function inside the summation sign in this solution is of exactly the same kind as the function S of Section 2.3 and that the resulting expression for the temperature is then the same as Equation (3.18). Exactly the same principle can be extended to a continuous distribution of point sources along a curve C in the thickness of the workpiece. Suppose that the power absorbed per unit length of the curve at a point on the curve with position vector $\mathbf{r}_0$ is $Q(\mathbf{r}_0)$. The increment to the temperature distribution in an infinite block caused by a short length $|d\mathbf{r}_0|$ of C is therefore

$$\frac{Q(\mathbf{r}_0)}{\lambda} \text{point}\left(\frac{Ux}{2\kappa}, \frac{Uy}{2\kappa}, \frac{U}{2\kappa}(z - z_0)\right) |d\mathbf{r}_0|.$$

In consequence, the increase in temperature in an infinite medium caused by the distribution of sources along C is

$$\frac{1}{\lambda} \int_{\mathbf{r}_0 \in C} Q(\mathbf{r}_0) \text{point}\left(\frac{Ux}{2\kappa}, \frac{Uy}{2\kappa}, \frac{U}{2\kappa}(z - z_0)\right) |d\mathbf{r}_0|$$

and in a semi-infinite medium with a nonconducting boundary at $z = 0$ is

$$S(x, y, z) = \frac{1}{\lambda} \int_{\mathbf{r}_0 \in C} Q(\mathbf{r}_0) \left\{ \text{point}\left(\frac{Ux}{2\kappa}, \frac{Uy}{2\kappa}, \frac{U}{2\kappa}(z - z_0)\right) + \right.$$
$$\left. + \text{point}\left(\frac{Ux}{2\kappa}, \frac{Uy}{2\kappa}, \frac{U}{2\kappa}(z + z_0)\right) \right\} |d\mathbf{r}_0|.$$

$$(6.4)$$

This function S has exactly the properties required of S in Section 3.3. Consequently, the temperature distribution in a plate caused by the variable line source consisting of the distribution of sources along C is again given by (3.18) with S given by (6.4).

The same principle can also be applied to distributions of point sources over a surface. Normally however the surface will be the surface of the workpiece and will have to be flat for the method to apply.

These ideas can be extended to regions with more complicated boundaries consisting of suitable straight-line sections or circles.

The idea has been applied to models of laser lap-welding of thin sheets of zinc-coated steel.[5] Two point sources were used, one located at the surface and one at the point of interface of the two sheets, together with a line source whose strength varied with position. The latter is constructed as a continuous distribution of point sources combined by integration. The boundary conditions were satisfied using the image technique just described. Figure 6.6 shows some examples. It will be seen that a degree of realism is possible with this representation that is not achievable with a single point source and a line source of constant strength. By making a guess at the forms for the individual elements it is possible to deduce likely distributions for the absorption at different depths in the weld. It should be noticed, however, that this is a difficult inverse problem and it is not easy to use it to draw accurate inferences about the power absorption characteristics of the process, helpful as such a procedure might be. All the same, in the long run it may be that use as a diagnostic tool proves to be the most useful application of the point and line source idea.

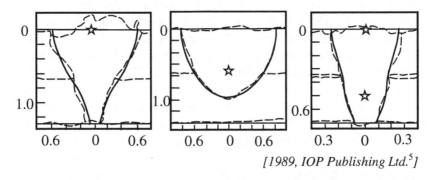

[1989, IOP Publishing Ltd.[5]]

Figure 6.6. Three examples of point and variable line source models of lap welds in zinc-coated steel sheets. The positions of the point sources are shown by stars. There is no line source in the second example, and the strength of the line source decreases with depth in the case of the other two. The observed profile is shown by a broken curve, and the calculated profile by the full curve.

---

[5] Akhter et al., 1989.

The entire problem of studying the welding process in principle requires the solution of systems of partial differential equations in a minimum of three separate regions (the solid, liquid, and vapor phases). There are interfaces between them that have to be determined as part of the solution of the problem (an example of what is known as a Stefan problem). It is conventional, therefore, to make assumptions about those parts of the process that are not the immediate subject of study. If what is required is a reasonably accurate knowledge of the temperature distribution far from the keyhole, as would be needed for example to determine the boundary of the weld itself, the heat affected zone, or the variation of metallurgical properties with position, it is an unwanted complication to have to perform detailed keyhole modeling in order to obtain the absorption profile of the keyhole. Such a process even involves the necessity of finding the keyhole size and shape. All this is necessary as well as the solution of Equation (6.1) in a region whose boundary is itself the keyhole wall that is determined by the heat balance condition. The process seems unduly cumbersome. What is required is information at a substantial distance from the keyhole, where the influence of its detailed shape and absorption characteristics might be assumed to be relatively unimportant.

It is in circumstances such as these that the strengths and weaknesses of the point and line source idea might profitably be exploited. The strengths are obvious in that the temperature distribution can be approximated without having to resort to obtaining special solutions of the equation of heat conduction. If the power distribution along the line source can be determined, the superposition principle can then be used to obtain the temperature far from the keyhole. The principal weakness is the way in which power absorbed in the walls of the keyhole is related to the radius of the keyhole when considered as part of the solid/liquid phases of the problem, and by absorption and transmission processes (inverse bremsstrahlung, Fresnel absorption, conduction, re-radiation, etc.) when the vapor phase is considered. There are two aspects to this: the inadequacy of these solutions near to the keyhole and the problem of determining the power absorbed from the laser beam by keyhole processes. Many different approaches have been adopted to overcome these problems.

Analytical approaches involve the use of formulae for specific absorption mechanisms; the Linking Intensity concept[6], Fresnel

---

[6] Dowden *et al.*, 1989.

absorption,[7] or alternatively empirical formulae can be employed to give the power absorbed per unit depth.[8] Whatever model is used for the power of the point sources and the absorption characteristics of the line sources, the result is a prediction for the temperature distribution in the workpiece. This can then be used as the basis for further calculation to obtain such quantities as

- the thermal history of individual points in the workpiece;
- the metallurgical properties (which can be deduced from the thermal histories);
- more accurate solutions for the temperature distribution, using the point and line source solution as the first stage of an iterative scheme, which might be analytical or, more usually, numerical;
- the consequences of varying the absorption models in order to test their validity and reliability;
- the distribution of thermal stress in the workpiece;
- the deformation of the workpiece that results from thermal stress.

It is characteristic of point and line source models that the nail-head is not as clearly defined as is often the case in practice. This is very clear when a comparison is made with the weld shown in Figure 6.2(b) above. Even the introduction of a variable line source model does not entirely solve the problem; see Figure 6.7, which shows that the weld of Figure 6.2(b) can be much more successfully represented. The shapes obtained are characteristically more like a wineglass than a nail-head, and therefore do not describe weld 6.2(a) at all well. Many reasons are possible, among them being the likelihood that there is strong convective motion in a large weld pool so that, in effect, power is being absorbed throughout the weld pool rather than at an isolated point. This can therefore be modeled by a volume distribution of point sources with, of course, the associated image system.

A related cause, at least in the case of the smaller nail-heads, is the possibility that the coupling of the beam into the hole in the workpiece is poor, so that a significant amount of power is absorbed over its upper surface. In that case a surface distribution of point sources could be used.

---

[7] See section 3.4.
[8] Ducharme et al., 1994b.

In the same way, the fact that the keyhole is usually curved rather than straight can be taken into account. The method is to integrate point sources with an appropriate source density along a curve in the interior of the material,[9] rather than along the straight line employed in the variable line source models described above. Again, an associated image system is required.[10]

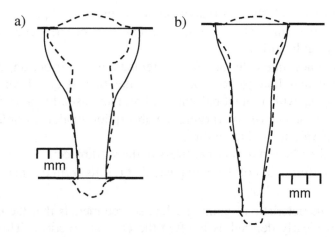

Figure 6.7. The welds of Figure 6.2 superimposed on a point and variable line source model. Observed profiles are shown as broken lines, and the calculated profiles as solid lines.

The applications of the idea of point and line sources can range from very simple models to a tool that can be used in the analysis, evaluation, and prediction of extremely complex problems of great industrial importance.[11]

The periodic point and line source solutions are given in Section 4.3 by Equations (4.12) to (4.14). The way in which the periodic time-dependant point source can be used to construct solutions for periodic surface heating, such as might occur in treatment with a pulsed laser, is described there. Clearly, the same mathematical strategy could be used with the time-dependent line source solution. Caution is needed since

---

[9] Kaplan, 1997.

[10] All these extensions are equivalent in mathematical terms to solving the problem by means of methods based on the use of an appropriate Green's function; see Morse and Feshbach, 1953.

[11] Ducharme et al., 1996.

the solution ignores the presence of a keyhole that may collapse and then be redrilled if the interval between pulses is sufficiently long. In that case a careful evaluation of the relevance of the approach is needed in any given application.

6.3   THE RELATION BETWEEN POWER ABSORBED AND KEYHOLE RADIUS

6.3.1 The line source as a model of the keyhole

The line source has proved to be very useful as a simple model of the temperature distribution in keyhole welding, but it has a number of drawbacks. For example, the temperature it predicts is above the vaporizing value sufficiently close to the axis, but it makes no allowance for that. It is possible, however, to use it to find an estimate of the keyhole radius. Such an estimate can be found by requiring the mean temperature on a circle of radius a to be the boiling temperature, T_B. From Equation (3.21) it means we require that

$$T_B = T_0 + \frac{Q}{2\pi\lambda}\frac{1}{2\pi}\int_{\theta=0}^{2\pi}\exp\left(\frac{Ua\cos\theta}{2\kappa}\right)K_0\left(\frac{Ua}{2\kappa}\right)d\theta .$$

But[12]

$$I_0\left(\frac{Ua}{2\kappa}\right) = \frac{1}{\pi}\int_{\theta=0}^{\pi}\exp\left(\frac{Ua\cos\theta}{2\kappa}\right)d\theta ,$$

and so the requirement is that

$$\tau I_0(a')K_0(a') = 1 \text{ where } \tau = \frac{Q}{2\pi\lambda(T_B - T_0)} \text{ and } a' = \frac{Ua}{2\kappa}. \quad (6.5)$$

Figure 6.8 gives a graph of a' as a function of $1/\tau$ on a logarithmic plot. As an illustration, 1 kW of laser power absorbed in 1 cm of stainless steel gives a value for τ of about 0.4, resulting in a value for a' of 0.93. Interestingly, over the range shown the relation is nearly linear;

---

[12] Abramowitz and Stegun, 1965, formula 9.6.16.

this is not a coincidence since the values of a' in the range are small enough for asymptotic approximations to be used for the Bessel functions. The approximations that are useful in this instance are[13]

$$K_0(s') = -\left\{\ln\tfrac{1}{2}a' + \gamma\right\} + O\!\left(a'^2 \ln\tfrac{1}{2}a'\right), \quad I_0(a') = 1 + O\!\left(a'^2\right),$$

where $\gamma = 0.57721566\ldots$ is the Euler-Mascheroni constant. Consequently Equation (6.8) can be approximated by

$$1/\tau = -\gamma - \ln\tfrac{1}{2}a' \quad \text{or} \quad a' = 2\exp(-\gamma - 1/\tau). \tag{6.6}$$

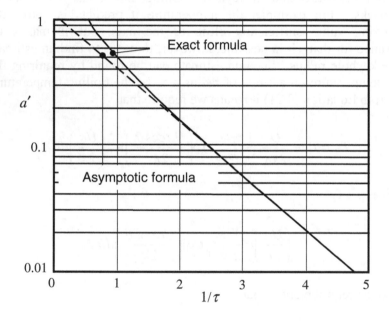

Figure 6.8. The exact and asymptotic relations between the dimensionless keyhole radius a' and the dimensionless absorbed power τ, as given by the line-source solution.

Approximations such as this are often useful, but it is important to remember the circumstances under which they are valid. Figure 6.8

<hr />

[13] Abramowitz and Stegun, 1965, Formulae 9.6.12 and 9.6.13.

shows both curves. From there it can be seen that the approximate relation is useful if τ is less than about 0.5 or a' less than about 0.2. For $\tau = 1$, the error in using the approximate relation is already about 20%.

Relationships such as these have to be used with an understanding of the context. Since they are a description of just a part of a much wider set of interrelations, it can be difficult, or even meaningless, to distinguish between cause and effect. The keyhole size and absorbed laser power are connected. The beam diameter is one of the main determining factors, but the relationship between it and the keyhole size is not necessarily straightforward, except at the keyhole mouth where (6.5) does not apply.

6.3.2 The Davis-Noller solution

One reason why the line source might be considered unsatisfactory as a model for the keyhole boundary is as follows. The high intensity of the laser beam incident on the keyhole wall forces the wall to have a radius that is closely connected to the beam radius. If the keyhole surface is taken to be the boiling isotherm, the line-source model could be used more accurately if that isotherm is identified; that, however, is not entirely a simple matter, as Figure 3.17 shows.

A similar but rather more sophisticated solution of the two-dimensional equation of heat conduction exists that gets around that particular problem at the expense of introducing some others. The basic requirement is that the temperature distribution $T(r,\theta)$ must satisfy the conditions that

$$T(a,\theta) = T_B \text{ and } T \to T_0 \text{ as } r \to \infty.$$

Here, r is the distance from the axis and θ is the angular coordinate in polar coordinates as shown in Figure 6.9. All quantities are assumed independent of the z coordinate. T must satisfy the equation of heat conduction.

To solve the problem, make the same substitution as in the discussion of the point and line sources given by Equation (3.2). Once again, S must satisfy Equation (3.3). This time, however, look for a solution for S that depends on both of the cylindrical polar co-ordinates

r and θ. Equation (3.3) then becomes[14]

$$\frac{\partial^2 S}{\partial r^2} + \frac{1}{r}\frac{\partial S}{\partial r} + \frac{1}{r^2}\frac{\partial^2 S}{\partial \theta^2} = \frac{U^2}{4\kappa^2}S. \tag{6.7}$$

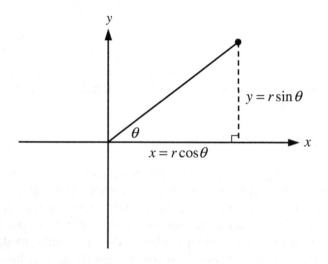

Figure 6.9. The polar coordinate system.

Look for solutions of the form

$$S = R_n(r)\cos n\theta \ \text{ where } n \text{ is an integer.}[15]$$

This particular form is suggested by the requirement that the temperature distribution should be symmetric about the plane $y = 0$ corresponding to $\theta = 0, \pi$, and the fact that the solution must have the same value at any point with coordinates $(r, \theta + 2\pi)$ as it does at (r, θ). The first of these requirements leads to the rejection of multiples of $\sin n\theta$, and the second to the requirement that n must be an integer. If this form is substituted into Equation (6.8) it is found that R_n must satisfy

---

[14] Kreyszig, 1993, Ch.11.
[15] Derived as an example of the solution of a partial differential equation by the method of separation of variables.

$$r^2 \frac{d^2 R_n}{dr^2} + r\frac{dR_n}{dr} - \left(\frac{U^2 r^2}{4\kappa^2} + n^2\right) R_n = 0 . \qquad (6.8)$$

This is once again an example of Bessel's equation, and its solutions are[16]

$$R_n = A_n \, \mathrm{I}_n\!\left(\frac{Ur}{2\kappa}\right) + B_n \, \mathrm{K}_n\!\left(\frac{Ur}{2\kappa}\right)$$

which will lead to contributions to T of the form

$$\exp\!\left(\frac{U}{2\kappa} r\cos\theta\right)\!\left[A_n \, \mathrm{I}_n\!\left(\frac{Ur}{2\kappa}\right) + B_n \, \mathrm{K}_n\!\left(\frac{Ur}{2\kappa}\right)\right]\cos n\theta .$$

The requirement that T must tend to zero at infinity in all directions means that all the coefficients A_n must be zero, leading to a composite solution for T obtained by adding together all of these possible individual solutions of the equation of heat conduction of the form

$$T = T_0 + \sum_{n=0}^{\infty} B_n \, \exp\!\left(\frac{U}{2\kappa} r\cos\theta\right)\mathrm{K}_n\!\left(\frac{Ur}{2\kappa}\right)\cos n\theta .$$

The boundary condition on the cylindrical keyhole wall then requires that

$$T_B - T_0 = \sum_{n=0}^{\infty} B_n \, \exp\!\left(\frac{U}{2\kappa} a\cos\theta\right)\mathrm{K}_n\!\left(\frac{Ua}{2\kappa}\right)\cos n\theta .$$

The problem then becomes that of finding the coefficients B_n. It is in fact relatively straightforward. Multiplying the condition by $\exp\!\left(-\dfrac{U}{2\kappa} a\cos\theta\right)\cos m\theta$ where m is an integer and integrate from 0 to 2π. Since

---

[16] Abramowitz and Stegun, 1965, Formula 9.6.1.

$$\int_0^{2\pi} \cos m\theta \cos n\theta \, d\theta = \begin{cases} 0 & n \neq m \\ \pi & n = m \neq 0 \\ 2\pi & n = m = 0. \end{cases}$$

It follows that

$$(T_B - T_0)\int_0^{2\pi} \exp\left(-\frac{Ua}{2\kappa}\cos\theta\right)\cos m\theta \, d\theta = \begin{cases} \pi B_m \, \mathrm{K}_m\left(\dfrac{Ua}{2\kappa}\right) & m \neq 0 \\ 2\pi B_0 \, \mathrm{K}_0\left(\dfrac{Ua}{2\kappa}\right) & m = 0. \end{cases}$$

But[17]

$$\int_0^{2\pi} \exp\left(-\frac{Ua}{2\kappa}\cos\theta\right)\cos m\theta \, d\theta = 2\pi(-1)^m \, \mathrm{I}_m\left(\frac{Ua}{2\kappa}\right),$$

so that

$$B_0 = (T_B - T_0)\frac{\mathrm{I}_0(Ua/2\kappa)}{\mathrm{K}_0(Ua/2\kappa)}$$

and

$$B_n = 2(T_B - T_0)(-1)^n \, \frac{\mathrm{I}_n(Ua/2\kappa)}{\mathrm{K}_n(Ua/2\kappa)} \quad \text{for } n > 0.$$

Hence

$$T = T_0 + (T_B - T_0)\exp\left(\frac{Ux}{2\kappa}\right) \times$$

$$\times\left\{\frac{\mathrm{I}_0(Ua/2\kappa)\mathrm{K}_0(Ur/2\kappa)}{\mathrm{K}_0(Ua/2\kappa)} + 2\sum_{n=1}^{\infty}(-1)^n \, \frac{\mathrm{I}_n(Ua/2\kappa)\mathrm{K}_n(Ur/2\kappa)}{\mathrm{K}_n(Ua/2\kappa)}\cos n\theta\right\}$$

where $x = r\cos\theta$.[18]

---

[17] Abramowitz and Stegun, 1965, Formula 9.6.19. Replace z by $-z$ and note from Formula 9.6.10 that when n is an integer, I_n is even if n is even, odd if n is odd.

[18] Abramowitz and Stegun, 1965, formula 9.6.19.

This is the Davis-Noller solution[19] for the temperature distribution around an isothermal cylinder.[20] From it, a revised relation can be obtained between the power of the line source and the radius of the keyhole. The flux of thermal energy per unit area across a surface with unit normal $\mathbf{n}$ is $\left(-\lambda\nabla T + \rho c_p T\mathbf{u}\right).\mathbf{n}$. Mass conservation and the fact that $r = a$ is an isotherm ensures that the second term, the convective contribution, has a net value of zero so that the power per unit depth into the workpiece is

$$Q = -\lambda a \int_{\theta=0}^{2\pi} \frac{\partial T}{\partial r}\bigg|_{r=a} d\theta = -\lambda a \left[\frac{d}{dr}\int_{\theta=0}^{2\pi} T\, d\theta\right]_{r=a}.$$

While it is possible to evaluate this integral directly it is simpler to notice that the heat flux into the workpiece across it must, in a steady problem, be equal to the heat flux out across a cylinder of arbitrarily large radius. Consequently,[21]

$$\int_{\theta=0}^{2\pi} \exp\left(\frac{Ur}{2\kappa}\cos\theta\right)\cos n\theta\, d\theta = 2\pi\, I_n\left(\frac{Ur}{2\kappa}\right)$$

so that

$$Q = -\frac{\lambda a U \pi}{\kappa}(T_B - T_0)\left\{\frac{I_0\left(I_0'K_0 + I_0 K_0'\right)}{K_0} + 2\sum_{n=1}^{\infty}(-1)^n \frac{I_n\left(I_n' K_n + I_n K_n'\right)}{K_n}\right\}$$

$$(6.9)$$

where all the Bessel functions are evaluated at $Ua/2\kappa$. However,[22]

$$2I_n' = I_{n-1} + I_{n+1}, \quad 2K_n' = -K_{n-1} - K_{n+1},$$

$$K_{n+1} I_n = \frac{2\kappa}{Ua} - K_n I_{n+1}, \quad K_{n-1} I_n = \frac{2\kappa}{Ua} - K_n I_{n-1},$$

---

[19] Davis, 1983; Noller, 1983.
[20] A similar solution exists for an isothermal sphere, and is obtained in much the same way; see Dowden et al., 1995b.
[21] Abramowitz and Stegun, 1965, Formula 9.6.19.
[22] Abramowitz and Stegun, 1965, Formulae 9.6.26 and 9.6.18.

so that

$$2(-1)^n \frac{I_n (I'_n K_n + I_n K'_n)}{K_n} = 2(-1)^n (I_{n-1} I_n + I_n I_{n+1}) - \frac{2\kappa(-1)^n}{Ua} \frac{I_n}{K_n}$$

and, in the same way,[23]

$$\frac{I_0 (I'_0 K_0 + I_0 K'_0)}{K_0} = 2 I_0 I_1 - \frac{2\kappa}{Ua} \frac{I_0}{K_0} .$$

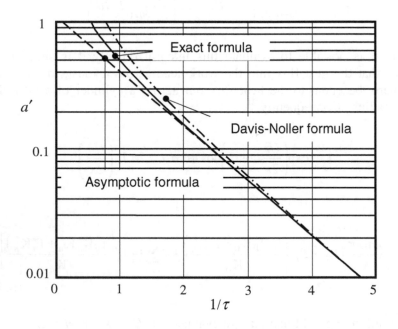

Figure 6.10. The relations between the formulae connecting power absorbed per unit length of the keyhole (τ in dimensionless form) and keyhole radius (a' in dimensionless form) given by the two Formulae (6.5) and (6.10) together with the asymptotic relation.

The terms involving products of the I Bessel functions sum to zero,[24] leaving only the ones with the ratios of I to K. Consequently,

---

[23] Abramowitz and Stegun, 1965, formulae 9.6.27 and 9.6.15.

[24] It is suggested that the first few terms should be written out explicitly to see the way in which it occurs.

$$Q = 2\pi\lambda(T_B - T_0)\,\mathrm{cyl}\!\left(\frac{Ua}{2\kappa}\right) \tag{6.10}$$

or

$$\tau = \mathrm{cyl}(a'), \tag{6.11}$$

where

$$\mathrm{cyl}(a') = \frac{1}{\tau}\frac{\mathrm{I}_0(r)}{\mathrm{K}_0(r)} + 2\sum_{n=1}^{\infty}(-1)^n\,\frac{\mathrm{I}_n(r)}{\mathrm{K}_n(r)}.$$

The two estimates obtained from (6.5) and (6.10) are shown for comparison in Figure 6.10, and it will be seen that there is little difference for sufficiently small values of a'. The asymptotic relation for small values of a' is applicable to both.

6.3.3 Other two-dimensional thermal models for the solid and liquid phases

The Davis-Noller solution tries to extend the line source idea to cover the case of a circular isothermal keyhole of nonzero radius. It suffers, however, from one serious departure from reality. The material elements all move in a straight line, and all those within a distance equal to the radius of the keyhole cross the boiling isotherm into the keyhole. They then cross back on the downstream side of the keyhole. While it is true that some material is vaporized and then escapes axially along the keyhole, it is not believed to recondense. A far better approximation would be to ignore the vaporization entirely and require the keyhole boundary to be a streamline as well as an isotherm.

Various approaches have been tried. For example, if the workpiece is regarded as entirely liquid and that this liquid is in irrotational flow around the keyhole, a solution has been found for the temperature distribution.[25] An alternative approach is to tackle the problem in the case of very slow relative motion. It is then possible to obtain an approximate solution that identifies the boundary between the solid and liquid phases and can test hypotheses about the importance of such

---

[25] Colla et al., 1994.

things as the latent heat of melting or the viscosity of the molten phase.[26]

A different use of the Davis-Noller solution is to use it to describe the temperature distribution in the solid phase when the solid/liquid boundary is on an isothermal circular cylinder.

6.4 THE LIQUID/VAPOR INTERFACE

A subject that has been investigated fairly extensively by means of simple models is the shape of the keyhole in deep-penetration welding. The subject is, however, by no means closed, and it is worth looking briefly at some of the individual considerations that have to be taken into account.

Important questions to ask are, What forces keep the keyhole open? How much energy is needed to do this? A complete attempt to answer such questions leads to a system of linked nonlinear partial differential equations. Order of magnitude calculations, however, suggest that the keyhole is kept open against surface tension forces mainly by "ablation pressure." Suppose the components of velocity normal to the interface and the density are u and ρ, with their values on the two sides distinguished by subscripts L and V, respectively, as indicated in Figure 6.11.

Conservation of mass requires that

$$\rho_L u_L = \rho_V u_V \equiv \dot{m}_A \qquad (6.12)$$

while the net momentum crossing the boundary per unit area from the liquid to the vapor side is

$$\rho_L u_L^2 - \rho_V u_V^2 = \dot{m}_A^2\left(\frac{1}{\rho_L} - \frac{1}{\rho_V}\right) \approx -\frac{\dot{m}_A^2}{\rho_V}. \qquad (6.13)$$

---

[26] Rogers, 1977; Dowden et al., 1983.

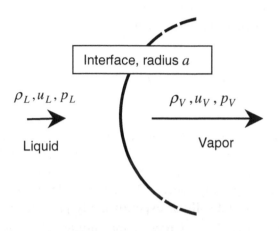

Figure 6.11. Schematic representation of material crossing a boundary at which vaporization occurs.

Use has been made of Equation (6.12). By Newton's Second Law such a momentum deficit has to be balanced by appropriate forces. There are two of these, provided by pressure and surface tension. If the keyhole has a radius a, viscous forces are ignored and the coefficient of surface tension for the molten metal is γ, then the sum of the forces per unit area at the boundary in the direction of the arrows in Figure 6.11 is

$$p_V - p_L - \frac{\gamma}{a} \approx -\frac{\gamma}{a} \qquad (6.14)$$

on the assumption that the pressure in the keyhole and in the liquid phase are both close to the ambient pressure.[27]   Equation (6.14) therefore gives the approximate result that

$$a\dot{m}_A^2 = \gamma p_V . \qquad (6.15)$$

In this equation a will need to be replaced by the reciprocal of the sum of the reciprocals of the principal radii of curvature if the shape of the

---

[27] This, however, must be regarded as contentious. It is possible to analyze the pressure at the keyhole wall in a number of different ways. Some approaches suggest that the pressure in the keyhole may differ substantially from the ambient pressure and the pressure in the liquid. See, for example, Kroos et al., 1993a,b; Gillies, 2000.

keyhole departs significantly from a circular cylinder. The total power per unit depth needed to produce such a rate of ablation in a circular keyhole is then

$$Q_A = 2\pi \left\{ c_{pS}(T_M - T_0) + c_{pL}(T_B - T_M) + L_M + L_V \right\} \sqrt{a \gamma \rho_V} .$$

(6.16)

Here, T is temperature with subscripts 0, M, and V indicating ambient conditions, melting, and boiling conditions, respectively, while L is the latent heat. Subscripts S and M refer to the solid and molten states, while T_B is the boiling temperature. The accurate determination of the vapor density ρ_V theoretically or experimentally presents considerable difficulties (see Section 7.3). Order of magnitude calculations can be made on this expression showing that, for typical beam sizes, the amount of power needed to keep the keyhole open is a fairly small part of the power absorbed. For this reason it tends to be ignored in simple models.

Something that is believed to have a profound effect on the shape of the weld pool is the variation of surface tension with temperature, resulting in what is known as the Marangoni effect or thermocapillary action. Once again, mathematically, this is a part of the boundary conditions. On the inside walls of the keyhole the temperature is at or near the boiling temperature of the metal, but on the surface of the weld pool away from direct incidence by the laser beam there can be a surface tension gradient. Its form is $\frac{d\gamma}{dT}[\nabla T - \mathbf{n}(\mathbf{n}.\nabla T)]$[28] and it is a tangential surface force per unit area. It has to be included in the tangential component of the momentum balance condition in the same way that the ordinary surface tension force has to be included in the normal component of the momentum balance.

The transfer of energy from the laser beam to the workpiece takes place through two main mechanisms. The first is direct absorption at the keyhole wall, often referred to as *Fresnel absorption*. See Equation (3.19). The second is a more complicated process that begins with *inverse bremsstrahlung absorption* in the ionized vapor. This process

---

[28] ∇T is the vector $\left(\frac{\partial T}{\partial x}, \frac{\partial T}{\partial y}, \frac{\partial T}{\partial z} \right)$ and $\mathbf{n}$ is a unit normal to the surface.

transfers energy from the laser to the plasma. It is then transferred through the plasma by a number of mechanisms, of which thermal conduction, reradiation, and absorption are a few, to the keyhole walls. It is there absorbed into the liquid phase of the workpiece, or else used to ablate the material necessary to keep the keyhole open.

A formula widely used for the reflection coefficient $\mathcal{R}$ in Fresnel absorption is given by Equation (3.19). It applies to circularly polarized light. It is probably appropriate for wavelengths equal to and greater than that of a CO_2 laser, but it is questionable how appropriate it is for shorter wavelengths.

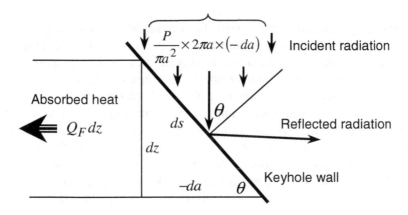

Figure 6.12. Geometry of the absorbing region at the wall of the keyhole.

If the keyhole is circular with radius $a(z)$, z being measured downwards from the surface of the workpiece, the power absorbed per unit depth of the keyhole by the Fresnel process Q_F is therefore obtained from the energy balance

$$Q_F \, dz = \frac{P(z)}{\pi a^2}\left[1 - \mathcal{R}(\theta)\right]\cos\theta \, 2\pi a \, ds \, .$$

Here, $\tan\theta = -\dfrac{dz}{da}$, $ds^2 = dz^2 + da^2$ and $P(z)$ is the total power at a cross-section of the keyhole perpendicular to the axis of the laser at depth z (see Figure 6.12). It is assumed here that the power is always uniformly distributed over the cross-section and that it is traveling parallel to the laser axis. This leads to the equation

$$Q_F = -[1 - \mathcal{R}(\theta)]P(z)\frac{2}{a}\frac{da}{dz}. \tag{6.17}$$

6.5 THE KEYHOLE

The keyhole is by far the most controversial part of the system to model effectively; a few of the considerations involved, however, are as follows.

A partially ionized vapor exists in the keyhole. The degree of ionization depends on the pressure and temperature in the keyhole if local thermodynamic equilibrium is assumed. Reasonable assumptions can be made about the pressure in the laser-generated keyhole so that, for local thermodynamic equilibrium, the keyhole temperature can be deduced. Before considering any nonequilibrium theories it is reasonable to study the system on the assumption that local thermodynamic equilibrium exists. This can be done using the Saha[29] equation that relates the number densities of the different species in the plasma to its temperature. The number densities are important since the absorption coefficient for inverse bremsstrahlung[30] depends strongly on their values.

Saha's equation and the absorption coefficient for inverse bremsstrahlung, with graphs of their temperature dependence, are discussed in more detail in Section 7.3.

A complication is that at an ablating boundary the distribution of particles cannot be Maxwellian since it is necessarily one-sided, that is, velocities can only be away from the liquid region, not into it. As a result, there is a very thin adjustment layer present, the Knudsen layer.[31]

For CO_2 laser light,[32] a substantial plasma is likely to occur in the keyhole (though much less so for CO laser light), and for assumed local

---

[29] Vicenti and Kruger, 1965; Eliezer et al., 1986, Ch. 7.
[30] Hughes, 1975, p.44.
[31] Finke and Simon, 1990; Finke et. al., 1990.
[32] Wavelength 10.6 μm.

thermodynamic equilibrium, the plasma formation for Nd:YAG laser light should be negligible. Experiment shows the presence of some ionized vapor when a Nd:YAG laser[33] is used, however.

Experimental measurements for keyhole plasma temperatures are very difficult to obtain because of the presence of a plasma plume above the keyhole; as a result, it is very difficult to access the keyhole directly. Measurements by reliable experimental groups tend to range from about 5000 K to 18000 K, suggesting that the assumption of local thermodynamic equilibrium in the keyhole plasma may be incorrect. It may be that a nonequilibrium rate process with other complicating effects is a more appropriate description.

In spite of all these reservations, however, simple models can provide useful information – it is just that it is as well to be aware that there are reservations that need to be made.

The electromagnetic propagation problem that determines the distribution of electromagnetic power in the keyhole poses some difficulties. The keyhole is often not many wavelengths in diameter, suggesting that, in principle at least, Maxwell's equations should be solved. The geometry is nonuniform and there is an inhomogeneous atmosphere, a problem that has not been successfully solved. It is more usual to treat the problem by the method of rays. The technique has been found to be useful[34] but normally requires elaborate computational procedures, as the rays are traveling in a variety of directions and are reflected as well as absorbed at the keyhole walls, with absorption taking place in the plasma.

A contribution to the absorption of the laser energy is that due to inverse bremsstrahlung, for which accurate modeling is difficult for the reasons given above. In outline, the laser power is absorbed in the partially ionized vapor in the laser-generated keyhole and transferred to the walls of the keyhole by thermal conduction processes. The laser power is related to the average value across the keyhole, and the mean intensity of radiation is averaged over all directions. This, in turn, is related to the energy absorbed in the plasma that acts as a source term in the heat conduction equation in which the thermal conductivity of the plasma is a function of temperature. The heat conduction equation can

---

[33] Wavelength 1.06 μm.
[34] Solana and Ocaña, 1997; Solana and Negro, 1997.

be solved to determine the temperature of the plasma in the keyhole. The concept of a linking intensity $\mathfrak{L}$ has been introduced[35] and has been found to be a useful though approximate concept. Originally, it was obtained in the case where the vapor in the keyhole is fully ionized, but the general idea can be seen as follows. Suppose the power absorbed by the workpiece per unit thickness by this mechanism is Q_B (W m^{-1}) and the laser power averaged at each cross-section of the keyhole is P (W). Then suppose there is an intensity of available incident power $\mathfrak{L}$ (W m^{-2}) at the keyhole wall that depends solely on these two quantities for given materials. Dimensional analysis then shows that

$$Q_B^2 = \mathfrak{L}\, P. \tag{6.18}$$

The value of $\mathfrak{L}$ also depends on the wavelength in use. As a generalization, it will be shown in Section 7.3 that if a number of reasonable assumptions are made, there is a relationship of the form

$$Q_B = Q(P, p)$$

where Q does not depend on the keyhole radius or the welding speed, but does depend on the mean values of the pressure and the total power at a given cross-section of the keyhole, as well as on numerous physical constants, properties of the materials involved, and the operating frequency of the laser.

As such, this relation, like Equation (6.17), constitutes a part of the boundary conditions on the liquid region. They cannot be considered in isolation and need to be considered in the context of the energy balance in the keyhole itself.

As an illustration of the way in which ideas such as these can be used to construct a simple model of the keyhole, consider the following. It is possible to write down two simple energy balance equations in the keyhole. These are

---

[35] Dowden et al., 1989. The value for the linking intensity in the case of a CO_2 laser at about 18000 K and atmospheric pressure obtained in this reference is given as about $3\frac{1}{2}$ kW cm^{-2}, an estimate that may well be too high. It falls off rapidly with increasing frequency.

$$\frac{dP}{dz} = -(Q_F + Q_B) \tag{6.19}$$

and

$$Q_F + Q_B = Q_A + Q_{Th}. \tag{6.20}$$

Equation (6.19) expresses the condition that the decrease in power in the keyhole P crossing a plane perpendicular to the axis at a given depth z is due partly to the direct absorption at the keyhole wall by mechanisms such as Fresnel absorption, and partly due to indirect transfer by processes such as inverse bremsstrahlung and thermal conduction in the plasma. Formulae (6.17) and (6.18) might be used for this purpose. Equation (6.20) states that this power goes either into the ablation of material from the keyhole wall Q_A or into the absorption of thermal energy in the workpiece, Q_{Th}. Q_A was discussed above and is frequently neglected. Q_{Th} has to be obtained from whatever solution is used for the solution of the equations of heat conduction, etc., in the liquid and solid regions. A simple way is to employ the line source model, a procedure that is not strictly accurate unless conditions are independent of z, but might be used as an approximation if they only vary slowly. If the approximation is accepted, then Q in Equation (3.21) is the quantity Q_{Th}. Assume that the keyhole wall is an isotherm, as will be at least approximately the case if the boiling is not too vigorous; it has already been noticed that this only needs to be sufficient for the ablation pressure to keep the keyhole open against surface tension. In that case, the keyhole wall is given by setting the right-hand side of (3.21) equal to T_V with $r = a$ and $x = \cos\theta$; this, then, has to be solved for a. It will be seen that the result is not exactly a circular keyhole. It is, however, possible to use an estimated average value for a, and the easiest way to obtain it is to average Equation (3.21) so that it is given by

$$T_B = T_0 + \frac{Q_{Th}}{2\pi\lambda} I_0\left(\frac{Ua}{2\kappa}\right) K_0\left(\frac{Ua}{2\kappa}\right). \tag{6.21}$$

See Equation (6.5). So, for example, a simple model can be obtained by neglecting Q_A, ignoring Q_B, and taking $\mathcal{R}$ to have a constant value $\overline{\mathcal{R}}$. Equations (6.17) and (6.19) together integrate to give

$$P = P_0 \left(\frac{a}{a_0}\right)^{2\left(1-\overline{\mathcal{R}}\right)} \qquad (6.22)$$

in which the subscript 0 indicates conditions at the top of the keyhole. Since Equation (6.20) simplifies to $Q_F = Q_{Th}$, and with Equation (6.17) both are equal to $-(1-\overline{\mathcal{R}})P\dfrac{2}{a}\dfrac{da}{dz}$. The following single differential equation for a is obtained,

$$-\left(1-\overline{\mathcal{R}}\right)P_0 \left(\frac{a}{a_0}\right)^{2\left(1-\overline{\mathcal{R}}\right)} \frac{2}{a}\frac{da}{dz} I_0\left(\frac{Ua}{2\kappa}\right) K_0\left(\frac{Ua}{2\kappa}\right) = 2\pi\lambda(T_B - T_0). \quad (6.23)$$

When this has been solved, P can then be obtained from (6.22). As a matter of mathematical technique, Equation (6.23) is best solved for z in terms of a rather than the other way around. Write

$$a' = \frac{U}{2\kappa}a \text{ and } z' = \frac{\pi\lambda(T_B - T_0)}{(1-\overline{\mathcal{R}})P_0}\left(\frac{Ua_0}{2\kappa}\right)^{2\left(1-\overline{\mathcal{R}}\right)} z,$$

from which it is possible to obtain the simple differential equation

$$\frac{dz'}{da'} = -\frac{I_0(a')\,K_0(a')}{a'^{2\overline{\mathcal{R}}-1}}. \qquad (6.24)$$

It must be solved with the boundary condition

$$a' = a_0' \equiv \frac{Ua_0}{2\kappa} \text{ at } z = 0.$$

The model has serious defects, as can be seen from the number of approximations introduced, and is not at all appropriate to study such things as maximum penetration depth or the conditions near the mouth of a very deep keyhole. It does, however, show what can be done using quite simple ideas.　Experimentation with the model, for example, shows how critical the value of the reflection coefficient is.

Figure 6.13 shows the profile given by this equation with a value of 0.85 for $\overline{\mathscr{R}}$.

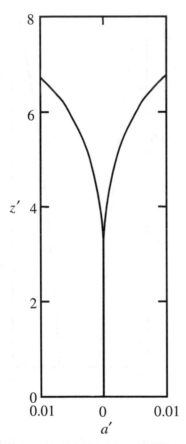

Figure 6.13. The profiles given by Equation (6.14) with a value of 0.85 for $\overline{\mathscr{R}}$. The vertical coordinate is measured upward from the maximum possible penetration depth obtainable from the model.

At first sight, the profile shown looks unrealistic, but consider the case where $a_0 = 0.1 \, \text{mm}$, $U = 1 \, \text{cm s}^{-1}$, $P_0 = 3 \, \text{kW}$, $h = 5 \, \text{mm}$. In that case $a_0' = 0.1$ and $h' = 1.5$. Figure 6.14 then shows only the relevant portion of the profile.

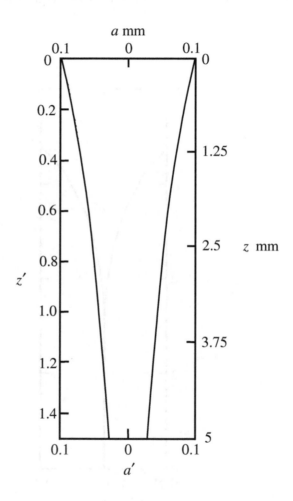

Figure 6.14. The profiles given by Equation (6.14) with a value of 0.85 for $\overline{\mathcal{R}}$. The case shown is when $a_0 = 0.1\,\text{mm}$, $U = 1\,\text{cm}\,\text{s}^{-1}$, $P_0 = 3\,\text{kW}$, $h = 5\,\text{mm}$. The dimensionless depth and radius are shown to the left and bottom of the figure, with the corresponding dimensional measurements to the right and top.

6.6 NUMERICAL EXAMPLES

Some simple calculations follow to illustrate the way in which the line source solution can be used for simple estimates of such things as the weld width. For example, if the numerical values for the constants of stainless steel given in Appendix 1 are used, Equation (3.21) can be used to estimate the width of the weld. Figure 6.15 shows the geometrical relationship between the weld width and the melting isotherm. It should be noticed that the maximum width of the molten region does not necessarily occur for the same value of x as the axis of the laser. At low translation speed the difference is not likely to be great. Figure 6.16 shows the numerical relationship in dimensionless form. Lengths are scaled with $2\kappa/U$ and

$$\tau = Q/2\pi\lambda(T_M - T_0). \tag{6.25}$$

So, for example, if a given keyhole weld absorbs 3 kW at a welding speed of $0.58\,\mathrm{cm\,s}^{-1}$, it is possible to estimate the width of the weld. The value of τ is approximately 0.5. The corresponding value for w from the graph in Figure 6.16 is 0.35 in dimensionless units, or equivalently 2.6 mm. This width is roughly the same as that of the lower part of the weld whose profile is shown in Figure 6.2b. Note that this is the width of the stem part of the weld, not the nail-head.

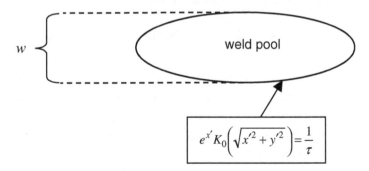

Figure 6.15. The geometrical relation between weld width and the melting isotherm for the line source expressed in dimensionless form.

The absorbed power is slightly different from the analysis above in terms of a point and a line source, although the total absorbed power in this instance is much the same in the two models. The presence of the point source makes a difference to a considerably greater depth than might be expected, a feature that may indicate a weakness in attempts to model the nail-head in terms of a point source.

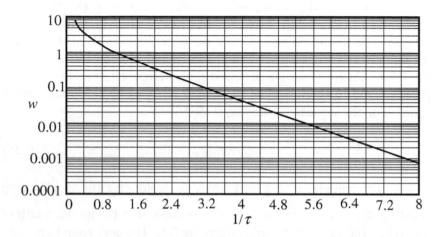

Figure 6.16. The numerical relation between weld width and the power absorbed per unit depth of the keyhole, τ, expressed in dimensionless form.

It is possible to use the approximation

$$\mathrm{K}_0(r) \approx -\gamma_E - \ln\!\left(\tfrac{1}{2} r\right),$$ valid for sufficiently small r,

where $\gamma_E = 0.5772...$ (the Euler-Mascheroni constant) to obtain the approximate relationship between τ and w. Assume that the weld pool is nearly circular, so approximately $w = 2r'$ and its center is close to the laser axis so that $x' \approx 0$. With this approximation for the Bessel function, the relationship

$$e^{x'} \mathrm{K}_0\!\left(\sqrt{x'^2 + y'^2}\right) = \frac{1}{\tau}$$

gives the approximate result

$$\log_{10} w = 0.3514 - 0.4343 / \tau \tag{6.26}$$

where use has been made of the equation $\ln x = \log_{10} x / \log_{10} e$. This formula gives a result that is almost identical to the part of Figure 6.25 that is a straight line. It is only at the left-hand end that there are significant differences.

It is possible to obtain a general form of the result as follows. The widest part of an isotherm occurs when $\dfrac{\partial y'}{\partial x'} = 0$ where y' and x' are related by the equation of the isotherm, $K_0(r')\exp x' = 1/\tau$ where $r' = \sqrt{x'^2 + y'^2}$. A general formula can be obtained as a result of this observation. Differentiation of the equation of the isotherm shows that at the value of x' corresponding to the greatest width,

$$\frac{x'}{r'} = \frac{K_0(r')}{K_1(r')}.$$

Consequently, the dimensionless weld width w is given by

$$w = 2\sqrt{r'^2 - x'^2}$$

where x' and r' are given as the solutions of

$$K_0(r')\exp x' = \frac{1}{\tau} \quad \text{and} \quad \frac{x'}{r'} = \frac{K_0(r')}{K_1(r')}.$$

It is this relationship that is plotted in Figure 6.15.

If the speed of translation is sufficiently small for the approximation given by (6.26) to be valid, the power absorbed per unit depth in the stem of the weld can be found. It is given in terms of the weld width W where $w = UW/2\kappa$ and the welding speed, U. It is obtained by substituting the definition of τ given by (6.25) into the approximate relation (6.26) and solving for Q. The result is the following equation,

$$Q = \frac{2.729\lambda(T_M - T_0)}{0.6524 - \log_{10}(UW/\kappa)}. \qquad (6.27)$$

This is only a valid approximation if UW is sufficiently small.

Equation (6.5) shows that the power absorbed per unit depth in a keyhole is given approximately by

$$Q = \frac{2\pi\lambda(T_V - T_0)}{I_0(\text{Pe})K_0(\text{Pe})} \text{ where } \text{Pe} = \frac{Ua}{2\kappa}.$$

Again, note that this applies to the stem part of the weld, not the nail-head. It is therefore possible to estimate the power per unit depth absorbed by a keyhole whose radius is known. Figure 6.17 shows a graph of $I_0(\text{Pe})K_0(\text{Pe})$.

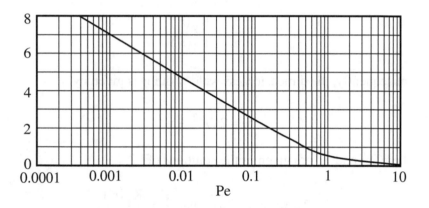

Figure 6.17. The function $I_0(\text{Pe})K_0(\text{Pe})$.

So, for example, if the keyhole has a diameter of 0.1 mm (or radius of 0.05 mm) at a welding speed of 0.58 cm s^{-1}, the value of Pe is 0.007, so p is about 5.2. Consequently, the power absorbed is about 2.3 kW m^{-1}.

The approximations $K_0(\text{Pe}) \approx -\gamma_E - \ln(\frac{1}{2}\text{Pe})$ and $I_0(\text{Pe}) \approx 1$ for small values of Pe can be used in the same way as before to obtain the approximate relation $I_0(\text{Pe})K_0(\text{Pe}) = 0.1159 - 2.303 \log_{10} \text{Pe}$. This approximate result can be combined with (6.27) to show that

$$W = 2 \left(\frac{1.1229\kappa}{U} \right)^{\frac{T_V - T_M}{T_V - T_0}} a^{\frac{T_M - T_0}{T_V - T_0}}$$

when these approximations are valid. The absence of Q from this is less surprising than at first appears, since a in the stem itself depends on Q and on all the absorption processes that occur higher up the keyhole. It can only be regarded as loosely related to the spot size of the laser.

Equation (6.6) could be used to express W in terms of Q instead, if desired. In that way, both the value of a and the approximate value of Q could be inferred from measurements of the weld width in the stem of the weld. The value at the surface will be wider as a result of the nail-head feature, and the theory would have to be extended to cover that part of the weld.

If the relationship between surface tension pressure and the rate of ablation given by Equation (6.15) is accepted, it is possible to estimate the power needed per unit depth to keep the keyhole open from equation (6.16). If, for example, the keyhole diameter is taken to be 0.1 mm, so that a is 0.05 mm, and the vapor density is taken to be 5×10^{-2} kg m^{-3}, Q_A has a value of about 1 kW cm^{-1}. This amount then has to be added to the other contributions to the total power needed for the welding process. It can be seen that although the figure is not big, it is not altogether a negligible fraction of the power absorbed by the workpiece in the form of heat.

FURTHER READING

The work piece

Avilov et al., 1996; Dilawari et al., 1978;
Dowden et al., 1987, 1991, 1995, 1998; Mara, 1974; Steen et al., 1988;
Swift-Hook and Gick, 1973;

The weld pool

Dowden et al., 1987, 1991, 1997; Ducharme et al., 1997;
Gratzke et al., 1992; Homann, 1936; Hopkins et al., 1994;
Kaplan, 1994; Kaye at al., 1983; Klemens, 1976; Kotecki et al., 1972;
Kroos et al, 1993; McLachlan, 1974; Miyazaki and Giedt, 1982;
Postacioğlu et al., 1989, 1991a, 1991b; Trappe et al., 1994.

The keyhole

Gouveia, 1994; Rykalin and Uglov, 1971.

CHAPTER 7

THE FLUID REGIONS IN KEYHOLE WELDING

7.1 FLOW IN THE WELD POOL

The molten metal in the weld pool is usually in vigorous fluid motion; there are several reasons why. One is that the molten metal has to flow *past* the keyhole or other surface depression rather than through it, as is implicitly assumed by the simpler models based on Equation (3.1). It is true that vaporization takes place on the keyhole wall, but only sufficient vaporization is needed to maintain the shape of the keyhole. Most of the fluid flows around it, although the flow can have a component parallel to the laser beam as well as perpendicular to it. Another driving mechanism is the variation of surface tension with temperature. Mathematically, the surface tension condition generates a boundary condition on the equations of fluid motion. The fact that there is a strong fluid motion means that the temperature distribution itself is affected. The fluid motion simply cannot be ignored in the way that it is in the simple models discussed in Chapter 6.

It is usual to describe the velocity field $\mathbf{u}$ in the liquid region by means of the equations of incompressible fluid dynamics, assuming that the density and kinematic viscosity are constant. These equations are

- The Navier-Stokes Equation (2.24)

$$\frac{\partial \mathbf{u}}{\partial t} + \mathbf{u}.\nabla \mathbf{u} = -\frac{1}{\rho}\nabla p + \nu \nabla^2 \mathbf{u} \qquad (7.1)$$

- and the Equation of Conservation of Mass (2.3)

$$\nabla.\mathbf{u} = 0. \qquad (7.2)$$

Furthermore, the motion is often turbulent. These equations, whether allowance is made for turbulence or not, generally have to be solved

191

numerically.   Nonetheless, there are some general considerations relating to boundary conditions on these equations that are important. The following points have to be borne in mind and need mathematical representation in a form appropriate to the given context. At a boundary between phases

(i) mass is conserved;

(ii) momentum is conserved in the absence of surface forces or modified appropriately in their presence (consider, for example, the effects of pressure or surface tension);

(iii) internal energy is conserved, making due allowance for latent heat effects.

These considerations apply whether or not there is a transfer of matter between the phases, although they tend to take much simpler forms when there is not.

Although some of molten metal is vaporized at the surface of the keyhole, most of it flows around. In general, the motion is very complicated, but some understanding of it can be obtained by studying simple analytical models. The most basic is to ignore the change of phase and treat the whole motion as if it were that of an inviscid fluid of constant density in irrotational motion, and to ignore vaporization at the

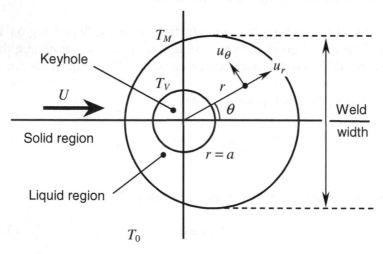

Figure 7.1. Schematic geometry of the keyhole and weld pool, with the Cartesian and polar coordinate systems employed.

keyhole wall. It is true that this is unsatisfactory on the level of detail, but it gives a useful approximation that can be employed to gain insight at a general level.

Assume that the fluid motion is independent of the z coordinate and that it has no component of velocity parallel to the keyhole's axis. Assume also that the keyhole is a circular cylinder of radius a. It is useful to employ cylindrical polar coordinates, (r, θ), centered on the keyhole axis. They are shown in Figure 7.1 in relation to the normal Cartesian axes.

From Equation (2.18) the velocity components of an irrotational flow in two dimensions are given by

$$(u,v) = \left(\frac{\partial \phi}{\partial x}, \frac{\partial \phi}{\partial y} \right) = \nabla \phi \text{ where } \nabla^2 \phi = 0 ; \qquad (7.3)$$

$\nabla^2 \phi$ is given in Cartesian coordinates by $\dfrac{\partial^2 \phi}{\partial x^2} + \dfrac{\partial^2 \phi}{\partial y^2}$, but since $x = r \cos \theta, x = r \cos \theta$, use of the chain rule[1] shows that

$$\nabla^2 \phi = \frac{\partial^2 \phi}{\partial r^2} + \frac{1}{r} \frac{\partial \phi}{\partial r} + \frac{1}{r^2} \frac{\partial^2 \phi}{\partial \theta^2} .$$

Since the two components of velocity in the polar coordinate system are given by

$$u_r = u \cos \theta + v \sin \theta = \frac{\partial \phi}{\partial r} \text{ and } u_\theta = -u \sin \theta + v \cos \theta = \frac{1}{r} \frac{\partial \phi}{\partial \theta} ,$$

and the boundary condition on the cylinder is simply $u_r(a, \theta) = 0$, it is easier to obtain a solution in terms of polar coordinates. There are two further conditions that have to be met. One is that far from the cylinder the velocity must be parallel to the x-axis and have magnitude U, and the other is that since there are no *a priori* reasons for expecting asymmetry, the solution must be symmetric in the x-axis. The first of

---

[1] Kreyszig, 1993, Ch.8.

these conditions means that, for large values of r, $\phi \sim Ux = Ur\cos\theta$ so that $(u,v) \sim (U,0)$. The second is consistent with the trial solution $\phi = f(r)\cos\theta$ inspired by the asymptotic conditions. Substitution of it into the polar coordinate form of $\nabla^2\phi = 0$,

$$\frac{\partial^2\phi}{\partial r^2} + \frac{1}{r}\frac{\partial\phi}{\partial r} + \frac{1}{r^2}\frac{\partial^2\phi}{\partial\theta^2} = 0,$$

shows that $f(r)$ must have the form $\phi = Ar + B/r$. The asymptotic form requires that $A = U$. The radial velocity is then given by

$$u_r = \frac{\partial\phi}{\partial r} = \left(U - \frac{B}{r^2}\right)\cos\theta.$$

The requirement that $u_r(a,\theta) = 0$ then shows that $B = Ua^2$ and, consequently,

$$\phi = U\left(r + \frac{a^2}{r}\right)\cos\theta.$$

The velocity components, in either Cartesian or polar form, can then be obtained by differentiation.

If all lengths are scaled with a, and velocities with U, a, and U disappear completely from the problem, any times that may be calculated are then scaled with a/U, and the velocity potential ϕ is scaled with Ua. This is equivalent to setting $a = U = 1$ in the formulation of the problem, but results in no loss of generality, so the following discussion will be presented as if this had been done for notational simplicity.

There are a number of ways of representing a fluid flow graphically. For example, one can plot the particle paths; these have to be obtained by solving Equation (4.20), which is often not a simple matter. An alternative is to construct the *stream lines* of the flow. These are a family of lines that are everywhere instantaneously parallel to the velocity vector. In steady flow, the stream lines and the particle paths are the same, but the stream lines are frequently much easier to obtain and tend to be preferred as a result. In unsteady flow, though, the two are not the same. Yet another way of representing the flow is by means of a vector plot in which vector line elements are drawn at the vertices of a mesh of points in the flow domain. Figure 7.2 is a vector plot of the velocity for the particular problem under discussion.

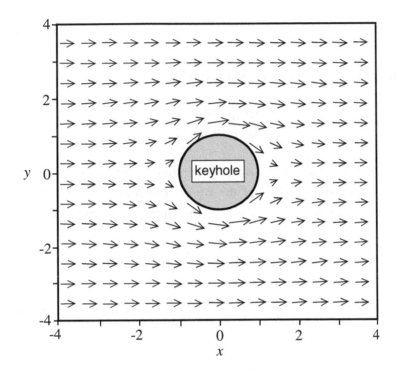

Figure 7.2. Vector plot of the velocity vector for irrotational flow past a cylinder of unit radius.

The stream lines can be found by integrating the equation

$$\frac{dx}{u} = \frac{dy}{v},$$

(7.4)

but since

$$u = \frac{\partial \phi}{\partial x}, v = \frac{\partial \phi}{\partial y} \text{ and } \frac{\partial^2 \phi}{\partial x^2} + \frac{\partial^2 \phi}{\partial y^2} = 0 ,$$

it follows that

$$\frac{\partial u}{\partial x} + \frac{\partial v}{\partial y} = 0,$$

and so there is a function ψ, called the stream function, such that

$$u = \frac{\partial \psi}{\partial y}, v = -\frac{\partial \psi}{\partial x} \quad .$$

Equation (7.4) can therefore be rewritten as

$$\frac{\partial \psi}{\partial x} dx + \frac{\partial \psi}{\partial y} dy = 0,$$

an expression which is the exact differential of ψ; the stream lines are therefore just the contours of the stream function given by $\psi = C$ for different values of C. Here the velocity potential ϕ is given by

$$\phi = x + \frac{x}{x^2 + y^2},$$

and by differentiation it is simple to verify that the corresponding stream function is

$$\psi = y - \frac{y}{x^2 + y^2}.$$

The stream lines corresponding to integer values of ψ are shown in Figure 7.3. Notice that the surface of the cylindrical keyhole is itself part of the stream line $\psi = 0$. That is to be expected since it was required that no liquid should cross the surface, so the flow there has to be tangential to it, and the stream lines are parallel to the velocity by definition.

Consider now the particle paths. Although these are the same as the stream lines in steady flow, the stream lines can give no information about the relative history of material elements. That can only be found by solving Equation (4.20).

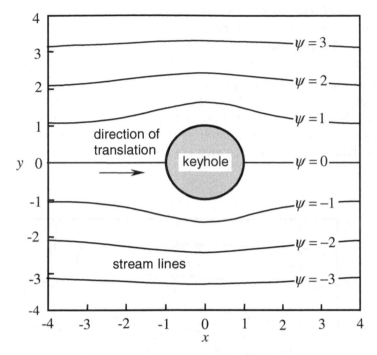

Figure 7.3.   The stream lines of steady flow past a cylinder of unit radius.

The equations for the Lagrangian coordinates $\xi(a,b,t)$ and $\eta(a,b,t)$ are therefore

$$\frac{\partial \xi}{\partial t} = 1 + \frac{\eta^2 - \xi^2}{\left(\xi^2 + \eta^2\right)^2}$$

$$\frac{\partial \eta}{\partial t} = -\frac{2\xi\eta}{\left(\xi^2 + \eta^2\right)^2}$$

with

$$\xi(0) = a, \quad \eta(0) = b.$$

Even for such a simple example as this, one is forced to solve the equations numerically, and of course such a procedure gives the stream lines. What is much more interesting, however, is to compare the relative positions of material elements at different distances from the line of symmetry $y = 0$. Figure 7.4 is the same as Figure 7.3, but superimposed on it are curves showing the positions at times $t = 3$ and $t = 6$ of elements that were initially at time $t = 0$ in a straight line at $x = -3$. The presence of the cylindrical keyhole has the effect of displacing fluid elements close to the plane $y = 0$ backwards relative to elements that were initially parallel to them, but further away from the plane of symmetry. The closer they are to this plane, the greater the effect, although one would not necessarily expect to see such extremes as this in practice, as the motion is in reality far more complicated. Qualitatively, however, this is what is found in practice.[2]

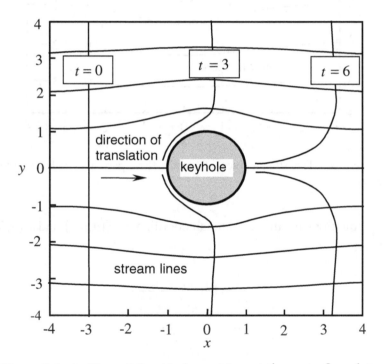

Figure 7.4. As Figure 7.3, with the positions at times $t = 3$ and $t = 6$ of material elements initially on the line $x = -3$.

The first weakness of solutions of this kind is that they do not distinguish between the solid and the molten phase. The difference may

---

[2]See, for example, Basalaeva and Bashenko, 1977, in the case of electron beam welding.

not be big but it needs investigation. One would expect to see stream lines more like those shown schematically in Figure 7.5 than those seen in Figure 7.3.

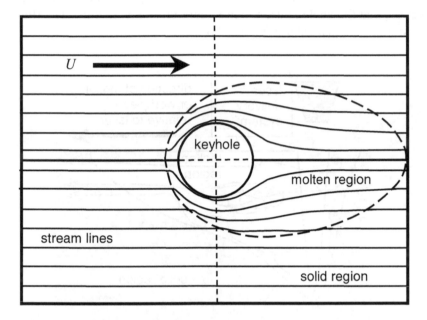

Figure 7.5. Schematic diagram of the stream lines expected in a two-dimensional motion in which the workpiece melts and resolidifies.

Many analyses have been made of varying degrees of complexity. It is, for example, possible to analyze the problem at low Péclet numbers,[3] where the keyhole and weld pool are both essentially circular in section, but with the weld pool slightly displaced. It is possible to include such features as the latent heat of melting and investigate the level of significance it has in determining the width of the weld.

---

[3] Dowden, 1983.

7.2 INTERACTION OF MOTION IN THE WELD POOL WITH THE KEYHOLE

It has been known for a long time[4] that there are flows in the weld pool parallel to the axis of the laser as well as perpendicular to it, and, indeed, that the keyhole itself is in violent unsteady motion. The flow is illustrated schematically in Figure 7.6.

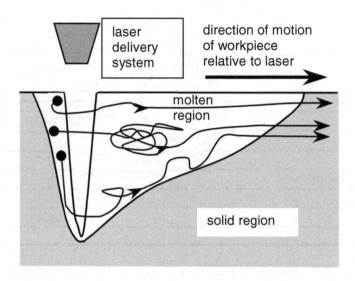

Figure 7.6. Schematic diagram showing typical particle paths in the weld pool.

It cannot be said that the cause is well understood, and although there have been theoretical investigations of it,[5] it is often ignored in models of keyhole welding. This is in contrast to the position in models of conventional welding techniques where it is usual to take proper account of it. Part of the problem is that it is not easy to identify the prime cause. There are almost certainly several different effects that contribute, but one possibility is convection induced by variation in the surface tension with temperature on the surface of the weld pool. The effect is known as *thermocapillary flow* or the *Marangoni effect*. The size of the effect can make a considerable difference in the

---

[4] Arata and Miyamoto, 1978; Arata,1987; Matsunawa, 2000.
[5] Postacioğlu et al., 1987.

characteristics of conventional welds.[6] Even the direction of the flow can depend critically on whether the gradient of surface tension is positive or negative, a difference that can be affected by surface-active impurities such as sulfur. Experimental and theoretical evidence lends support to the idea that differences in surface properties of melts account for variable weld penetration in conventional welding processes.[7]

The effects of thermocapillary flow have been investigated theoretically,[8] but the evidence that it is the main cause of flow parallel to the laser beam remains inconclusive, and other causes can be postulated.[9] The matter is not of trivial importance and is directly relevant to thermal modeling, since fluid convection in the weld pool will affect the way in which heat is redistributed within it. The width and depth of the resulting weld are therefore affected.

In a careful analysis of the problem, Matsunawa[9] listed four mechanisms that are known to induce circulation in the weld pool in arc welding, namely electromagnetic forces, buoyancy forces, surface tension, and aerodynamic drag at the surface of the workpiece. It should be remembered that a keyhole is not a normal feature of arc welding. It is known that bead formation in laser welding can be greatly influenced by an external magnetic field,[10] but in general electromagnetic effects are not important. The other causes remain as possibilities however. Matsunawa's list includes two other possibilities that can occur in keyhole welding that are not normally applicable in arc welding. These are the existence of humps that form on the keyhole wall[11] so that metal flow may be induced by the recoil force of evaporation, and liquid flow induced by the metallic vapor jet in the keyhole.

This last mechanism is clearly plausible, but is no easier to quantify than some of the others. In order to illustrate the techniques of theoretical investigation, a simple model will be constructed here, in the course of which it will become apparent that flow parallel to the axis of

---

[6] Heiple and Roper, 1982.
[7] Keene, 1988; the evidence was reassessed by Ishazaki, 1989.
[8] Postacioğlu et al., 1991a.
[9] Matsunawa, 2000.
[10] Kern et al., 2000.
[11] Matsunawa and Semak, 1997.

the laser is an essential consequence of the existence of the keyhole. There is no attempt at detailed numerical agreement; the purpose of the investigation is to identify underlying mechanisms, and qualitative agreement is the most that can be hoped for. Detailed numerical comparisons would require quite elaborate numerical simulation using the techniques of computational fluid dynamics.

The main problem in the construction of a model of something like this is how much to simplify. The keyhole and the molten region must be included, and it seems likely that the presence of the boundary represented by the solid part of the workpiece provides a constraint that may influence the pattern of flow. A large melt pool may result in a less tight circulation than a small one, making the effects of this flow less obvious. On the other hand, the thermal conditions that determine the size of the weld pool are not central to the problem under investigation, even though they are, in fact, consequences of the solution. Similarly, the thermal transfer mechanisms that determine the radius of the keyhole are not central, only its radius, so these two features can be regarded as given in advance. It would make analytical investigation easier if the two boundaries were regarded as concentric circular cylinders. The fact that they are not will accentuate the differences between the region in front of the keyhole and the region behind, but will not affect the qualitative features, so there is no obvious advantage in the very considerable complication that would result from including it. Suppose, therefore, that the keyhole is on the boundary $r = a$, the melting boundary is at $r = b$, and the workpiece occupies the region $0 \le z \le 2h$.

The central feature is the keyhole, which has to be kept open principally by the back-pressure of material ablating at its surface (see figure 7.7).

It is clear that the conditions at the interface between the liquid region and the vapor in the keyhole or the surrounding atmosphere at the surface of the molten portion of the workpiece is crucial. It is therefore necessary to look more closely at the fluid dynamics interface conditions discussed in Section 2.2. Since it is known that the boundary is not steady but is in vigorous motion, and is probably best described as being turbulent, there is nothing to be lost and perhaps quite a lot to be gained by considering the full time-dependent version of the conditions.

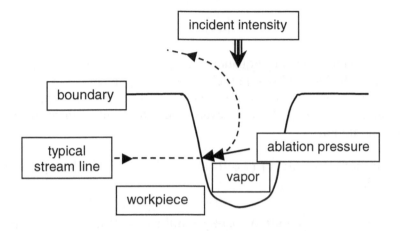

Figure 7.7. Ablation at the keyhole wall.

The first of these is given by Equation (2.48). If $\dot{m}$ is the mass flow rate per unit area normal to the interface, then

$$\dot{m} = \left[\rho(\mathbf{u}.\mathbf{n} - \dot{n})\right]_V = \left[\rho(\mathbf{u}.\mathbf{n} - \dot{n})\right]_L \qquad (7.5)$$

where $\mathbf{n}$ is the normal directed from the liquid region to the vapor and $\dot{n}$ is the velocity of the interface in the direction of its normal. The subscripts L and V indicate the liquid and vapor regions, respectively, while $[X]_L^V = X_V - X_L$. If $\mathbf{t}$ is any vector tangential to the surface, notice that

$$[\dot{m}]_L^V = [\dot{n}]_L^V = 0 \text{ and } [\mathbf{n}]_L^V = [\mathbf{t}]_L^V = 0.$$

The next fluid dynamic boundary condition is Equation (2.56), which states that

$$\left[\rho\mathbf{u}(\mathbf{u}.\mathbf{n} - \dot{n}) + p\mathbf{n} - \underline{\underline{\mathbf{d}}}.\mathbf{n}\right]_L^V = \mathbf{n}(\gamma\nabla.\mathbf{n}) + \{\mathbf{n}(\mathbf{n}.\nabla\gamma) - \nabla\gamma\} \qquad (7.6)$$

where $\underline{\underline{\mathbf{d}}}$ is the deviatoric stress tensor given in terms of the viscosity and the velocity gradients. See equation (2.21). The final condition is the no-slip condition (2.58),

$$[\mathbf{u.t}]_L^V = 0. \tag{7.7}$$

The discussion following Equation (2.58) shows that this applies when material melts into an inviscid region and the argument can be extended to the case of a viscous fluid crossing into an inviscid region. Conditions (7.5) to (7.7) apply even when the fluid is in unsteady or turbulent motion.

If the scalar product of (7.6) is taken with $\mathbf{n}$, the following result is obtained,

$$\left[\rho\mathbf{u.n}(\mathbf{u.n} - \dot{n}) + p - \mathbf{n}.\underline{\underline{d}}.\mathbf{n}\right]_L^V = \gamma\nabla.\mathbf{n}. \tag{7.8}$$

Use of (7.5) shows that the condition can also be written

$$\dot{m}(\dot{m} + \dot{n})\left[\frac{1}{\rho}\right]_L^V + \left[p - \mathbf{n}.\underline{\underline{d}}.\mathbf{n}\right]_L^V = \gamma\nabla.\mathbf{n}. \tag{7.9}$$

Essentially, Equation (7.9) relates the surface tension pressure, the term on the right, to the mass flux, the first term on the left. The viscous stress term is probably small compared to either of those. The term in $[p]_L^V$ is determined by the fluid dynamic flows in each of the two regions and may or may not be comparable to the surface tension pressure term.

Now take the scalar product of (7.6) with $\mathbf{t}$, use (7.7), and remember that $\mathbf{n.t} = 0$; then

$$\left[\mathbf{t}.\underline{\underline{d}}.\mathbf{n}\right]_L^V = \mathbf{t}.\nabla\gamma. \tag{7.10}$$

Thus, there are two conditions on the tangential components of stress and velocity given by (7.7) and (7.10). These conditions hold whether the flow is steady or unsteady. For discussions of turbulent flow they could, for example, be replaced by their averages, although it is then necessary to remember that $\mathbf{t}$ and $\mathbf{n}$ are also fluctuating quantities, as well as $\mathbf{u}$ and $\underline{\underline{p}}$. They show quite clearly that there are three things involved in the generation of tangential (and hence axial) flow in the weld pool. One is the no-slip condition (7.7), although this does not in itself imply the existence of tangential motion. Equation (7.10) shows that existence of a surface tension gradient is a powerful generator of tangential flow when taken in conjunction with viscous forces. That, however, does not necessarily lead to axial flow in the liquid region adjacent to the keyhole, since the walls of the keyhole are likely to be at or close to isothermal conditions. In consequence, the surface tension gradient is likely to be very small. That leaves the tangential condition in the very simple form

$$\left[\mathbf{t}.\underline{\underline{\mathbf{d}}}.\mathbf{n}\right]_L^V = 0. \tag{7.11}$$

This condition (7.11) makes it clear that the primary driving force is the transmission of tangential motion by the viscous forces in the two regions, the keyhole and the weld pool. To ignore viscous forces in the keyhole, for example, would lead to a zero-tangential-stress condition. That could be compatible with a zero-velocity condition at the keyhole wall, which, in turn, would lead to no axial flow, contrary to the experimental evidence.

A full solution of the problem involves the solution of linked fluid dynamical problems in each of the liquid and vapor regions. The two are connected by the boundary conditions (7.7), (7.9), and (7.11) and are beyond the scope of this book. It is, however, worth a partial investigation to see what the implications of the analysis might be, and whether they could be compatible with the evidence. It is much simpler, therefore, to separate the problem into two part, and investigate them separately, so far as that is possible.

Neither region is simple if it is true that the motions in them are fundamentally turbulent, but if the intention is a qualitative investigation, not a quantitative one, there are some simplifications possible. A simplifying assumption is to assume that the keyhole penetrates right through the workpiece and that the mean geometry is effectively cylindrical (see figure 7.8).

One is forced to the conclusion that the very existence of the keyhole necessarily leads to axial motion in the weld pool that may be very strong. It can be so strong, in fact, that the[12] keyhole could well be in a state of turbulent motion.

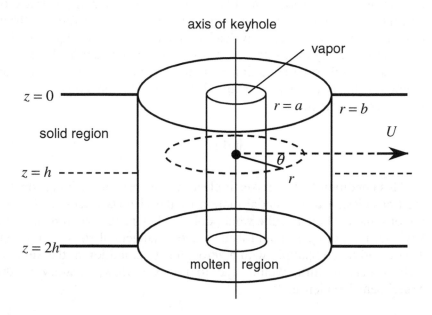

Figure 7.8. The coordinate system and boundaries in the keyhole and weld pool.

Experimental observations tend to confirm the supposition. A fact that is almost certainly related is that the keyhole is not a simple feature, stationary relative to the laser beam, but a twisting, writhing object whose position is strongly time-dependent. The same observations show, most remarkably and perhaps counter to initial expectations, that material close to the keyhole boundary moves down and away from the keyhole in blind keyholes. The experiments also note differences between behavior at the front and the rear of the weld pool, but this model relies on symmetry for the purposes of simplification, and so cannot be expected to imitate all such differences. It is easy to see how a tendency such as this, with flow in the weld pool towards the center of the workpiece parallel to the laser beam, can lead to the formation of pores. The same experimentalists found that pores could be avoided if the welding took place in a vacuum. In that case, flow in the weld pool

---

[12] Matsunawa, 2000; Katayama et al., 2000.

parallel to the laser beam took place in the same direction as the ejected vapor of the plume.

The magnitude of the fluid motion that can be induced by such effects has significant consequences for thermal modeling, although they are not easy to cope with, either analytically or computationally. Nonetheless, it is important to be aware of the problem. The reason is that if the fluid velocities are high relative to the speed of welding, the convection terms in the equation of heat conduction (2.6) result in strong mixing of fluid elements of different temperatures. If, in addition, the fluid motion is turbulent, these mixing effects can be very widely distributed throughout the molten region, amplifying the effect of the thermal conductivity very considerably. Exactly the same argument applies to the viscosity of the fluid whose effect is also increased, although the relation between the two different mechanisms is not straightforward.

Because of the way that turbulence can amplify the effects of conductivity and viscosity, and by analogy with the statistical dynamical derivations of the equations of conduction and viscous fluid motion, a simple way of taking account of turbulence is to introduce the concepts of "eddy viscosity" and "eddy conductivity." In the most basic form of such approximations, the ordinary viscosity μ is replaced by a constant *eddy viscosity* μ_E, and the thermal conductivity λ by a constant *eddy conductivity* λ_E. While it must be recognized that these are very simple approximations, and that there are no simple rules for determining appropriate values for them, they are approximations that can be used to test qualitative hypotheses or produce qualitative models as guides prior to more detailed numerical modeling. An excellent example of the latter procedure is provided by their use in early models of the Gulf Stream.[13]

Introduce eddy viscosity and conductivity coefficients to describe the mean effect of the complicated turbulent motions of the keyhole. Assume that they are sufficiently large that the inertial terms in the equations can be ignored in the region of molten metal.

To decouple the two regions, consider a simple description of the liquid region in which the normal component of velocity at the keyhole

---

[13] Stommel, 1965.

wall $r = a$ is taken to be independent of z and directed radially inward. This is an approximation to condition (7.9).

For the moment, ignore condition (7.7), which is part of the link between the motion in the weld pool and the flow in the keyhole.

Condition (7.11), then, has two parts. If **t** is in the azimuthal direction, that is to say, in the direction of increasing θ, there is unlikely to be a significant contribution in that direction from motion in the keyhole. The corresponding component $e_{r\theta}$ of the rate of strain tensor in the liquid region will then be effectively zero. The axial component however will not be zero, being given by (7.11) with **t** approximately $\hat{z}$ and **n** approximately $\hat{r}$. Approximately at least, the stress from the vapor in the keyhole may be expected to be an odd function about the central plane $z = h$.

The interface conditions on the three velocity components at the keyhole wall are then of the form

$$u_r^L(a, z) = -u_0$$

$$\left.\frac{\partial u_r^L}{\partial \theta}\right|_{r=a} - u_\theta^L(a, z) + a \left.\frac{\partial u_\theta^L}{\partial r}\right|_{r=a} = 0 \qquad (7.12)$$

$$\left.\frac{\partial u_z^L}{\partial r}\right|_{r=a} = -\frac{\tau_0}{\mu_E}\left(\frac{h-z}{a}\right).$$

The first of these is from (7.9), while the second and third are consequences of the requirement of specified (eddy) stresses. Thus the second is the θ component of the tangential surface stress condition. See Equations (2.22) with (2.5).[14] The third comes from the axial component of the tangential surface stress condition in the z-direction; $\tau_0(h - z)/a$ is the viscous stress caused by the vapor in the keyhole, and τ_0 can have either sign. It constitutes the second part of the linking conditions between the keyhole and the weld pool. It will be seen that if $\tau_0 > 0$, it implies that the axial velocity *decreases* immediately next to the keyhole wall, leading to an expectation of flow towards the

---

[14]See Landau and Lifshitz, 1959b, p.51, or Batchelor, 1967, p.602, for components of the stress tensor in cylindrical polars.

central plane. This is in agreement with the experimental evidence for welding at atmospheric pressure, with the flow reversed if $\tau_0 < 0$ as occurs if the welding is performed in a vacuum. The result depends on the fact that material is changing from the liquid to the vapor phase. In the latter state it has a high axial velocity outward from the keyhole, whereas in the liquid phase it has a much smaller velocity.

There are similarly three boundary conditions to be satisfied at the solid/liquid interface at $r = b$; they are simply conditions on the continuity of velocity at this boundary if the density of the two phases is taken to be the same. In connection with these and the first of the conditions (7.12), it is clear that there is a problem to be faced if the model is to remain simple. There is a loss of mass from the keyhole so that either the resolidified material does not fill the whole of the region $0 < z < 2h$, or else there is deformation of the solid material reducing the overall width of the workpiece – or, indeed, a combination of both. In either case, the simplicity of the geometry is lost, leading to increased difficulties in the solution of the problem. These difficulties are not central to the original purpose of the model, which was to look at vigorous fluid motion in the axial direction in the keyhole.

Since the radial velocities given by (7.12) are at most comparable to the translation speed U at the keyhole wall, the equivalent mass influx at the molten boundary would be reduced by a factor of a/b, a ratio that tends to be fairly small. The error, therefore, introduced by splitting the liquid velocity field into two parts, one of which is entirely radial, would be small. The appropriate solution is

$$u_r^L = -u_0 \frac{a}{r}, \ u_\theta^L = u_z^L = 0. \tag{7.13}$$

Hence, write

$$u_r^L = -u_0 \frac{a}{r} + v_r^L$$

where

$$v_r^L(a, z) = 0$$

$$\left.\frac{\partial v_r^L}{\partial \theta}\right|_{r=a} - u_\theta^L(a, z) + a\left.\frac{\partial u_\theta^L}{\partial r}\right|_{r=a} = 0 \tag{7.14}$$

$$\left.\frac{\partial u_z^L}{\partial r}\right|_{r=a} = -\frac{\tau_0}{\mu_E}\left(\frac{z-h}{a}\right)$$

and use the following boundary conditions at $r = b$,

$$v_r^L(b, z) = U \cos\theta$$
$$u_\theta^L(b, z) = -U \sin\theta \tag{7.15}$$
$$u_z^L(b, z) = 0.$$

There are also surface conditions to be imposed at the top and bottom of the melt pool. In principle, there are both velocity and stress conditions, and all of them together will determine the exact location of the molten surface. It is known in practice that it is at least approximately level with the surface of the solid part of the workpiece. Consequently, a reasonable approximation is to regard its location as being known, i.e., at $z = 0$ at the surface and at $z = 2h$ at the bottom. In that case there must be zero normal velocity there and zero tangential stress so that

$$u_z^L(r,0) = u_z^L(r,2h) = 0$$
$$\left.\frac{\partial u_\theta^L}{\partial z}\right|_{z=0} = \left.\frac{\partial u_\theta^L}{\partial z}\right|_{z=2h} = 0 \tag{7.16}$$
$$\left.\frac{\partial v_r^L}{\partial z}\right|_{z=0} = \left.\frac{\partial v_r^L}{\partial z}\right|_{z=2h} = 0.$$

The normal component of the stress condition could then be used to estimate the deviation of the actual surface from the assumed flat surface.

If the eddy viscosity is sufficiently high for the inertial terms to be ignored, the equations satisfied are Stokes' equations[15] and the equation of conservation of mass. In cylindrical polar coordinates[16] they are

---

[15] i.e., the Navier-Stokes equation (2.24) without the nonlinear inertial terms.
[16] Landau and Lifshitz, 1959b, p.51, or Batchelor, 1967, p.602.

$$\frac{1}{\mu_E}\frac{\partial p_L}{\partial r} = \frac{\partial^2 v_r^L}{\partial r^2} + \frac{1}{r^2}\frac{\partial^2 v_r^L}{\partial \theta^2} + \frac{\partial^2 v_r^L}{\partial z^2} + \frac{1}{r}\frac{\partial v_r^L}{\partial r} - \frac{2}{r^2}\frac{\partial u_\theta^L}{\partial \theta} - \frac{v_r^L}{r^2}$$

$$\frac{1}{r\mu_E}\frac{\partial p_L}{\partial \theta} = \frac{\partial^2 u_\theta^L}{\partial r^2} + \frac{1}{r^2}\frac{\partial^2 u_\theta^L}{\partial \theta^2} + \frac{\partial^2 u_\theta^L}{\partial z^2} + \frac{1}{r}\frac{\partial u_\theta^L}{\partial r} + \frac{2}{r^2}\frac{\partial v_r^L}{\partial \theta} - \frac{u_\theta^L}{r^2}$$

$$\frac{1}{\mu_E}\frac{\partial p_L}{\partial z} = \frac{\partial^2 u_z^L}{\partial r^2} + \frac{1}{r^2}\frac{\partial^2 u_z^L}{\partial \theta^2} + \frac{\partial^2 u_z^L}{\partial z^2} + \frac{1}{r}\frac{\partial u_z^L}{\partial r}$$

$$\frac{\partial v_r^L}{\partial r} + \frac{v_r^L}{r} + \frac{1}{r}\frac{\partial u_\theta^L}{\partial \theta} + \frac{\partial u_z^L}{\partial z} = 0.$$

$$(7.17)$$

This is a set of linear equations that can be solved by standard methods. It helps to notice that the pressure satisfies Laplace's equation, that u_z^L satisfies the biharmonic equation, and that the only inhomogeneities are in the boundary conditions (7.15), and are multiples of either $\sin\theta$ or $\cos\theta$. It is also helpful to separate the solution into two parts, one of which depends on θ; the other has no axial velocity and is axisymmetric. Use the symmetries of the problem to seek a solution in which

$$v_r^L = R(r)\cos\theta$$
$$u_\theta^L = \Theta(r)\sin\theta \qquad\qquad (7.18)$$
$$u_z^L = 0$$
$$p_L = \mu_E P(r)\cos\theta.$$

The boundary conditions (7.16) are then all identically satisfied and the part of conditions (7.15) that is independent of θ is satisfied if

$$R(b)=U, \quad \Theta(b)=-U. \qquad\qquad (7.19)$$

The part of the boundary conditions (7.14) independent of z are then satisfied if

$$R(a)=0$$
$$a\Theta'(a) = \Theta(a) + R(a). \qquad\qquad (7.20)$$

The required solution of (7.17) is the solution with these properties, of

$$P' = R'' + \frac{1}{r}R' - \frac{2}{r^2}(R+\Theta)$$

$$-\frac{1}{r}P = \Theta'' + \frac{1}{r}\Theta' - \frac{2}{r^2}(R+\Theta) \qquad (7.21)$$

$$R' + \frac{1}{r}(R+\Theta) = 0.$$

It follows that

$$R = \frac{2\ln(r/a)(a^4 + b^4) - b^2(r^2 - a^4/r^2)}{2\ln(b/a)(a^4 + b^4) - (b^4 - a^4)}U,$$

$$\Theta = -\frac{2\ln(r/a)(a^4 + b^4) + 2(a^4 + b^4) - b^2(3r^2 + a^4/r^2)}{2\ln(b/a)(a^4 + b^4) - (b^4 - a^4)}U,$$

$$P = -4\frac{\left[2rb^2 + (a^4 + b^4)/r\right]}{2\ln(b/a)(a^4 + b^4) - (b^4 - a^4)}U,$$

$$(7.22)$$

a result which can either be verified directly or obtained using a numerical algebra package.

Likewise, seek a z-dependent portion to the solution of the form

$$v_r^L = \cos\frac{n\pi z}{h}R_n(r)$$

$$u_\theta^L = 0$$

$$u_z^L = \sin\frac{n\pi z}{h}Z_n(r) \qquad (7.23)$$

$$p_L = \mu_E \cos\frac{n\pi z}{h}P_n(r).$$

where use has been made of the symmetry properties about $z = h$. The boundary conditions (7.16) are then all identically satisfied if n is an integer and the part of conditions (7.15) not dependent on θ are satisfied if

$$R_n(b) = Z_n(b) = 0, \quad n \geq 1. \qquad (7.24)$$

Boundary conditions (7.14) must then be satisfied by Fourier analysis, so that

$$R_n(a) = 0, \quad Z_n'(a) = -\frac{2\tau_0 h}{\pi a \mu_E} \frac{1}{n}. \tag{7.25}$$

Substitution of (7.23) into (7.17) shows that R_n, Z_n, and P_n must satisfy

$$P_n' = R_n'' + \frac{1}{r} R_n' - \left(\frac{n^2 \pi^2}{h^2} + \frac{1}{r^2} \right) R_n$$

$$-\frac{n\pi}{h} P_n = Z_n'' + \frac{1}{r} Z_n' - \frac{n^2 \pi^2}{h^2} Z_n \tag{7.26}$$

$$R_n' + \frac{1}{r} R_n + \frac{n\pi}{h} Z_n = 0.$$

Clearly, Bessel functions are likely to be involved, and from the fact that the pressure is necessarily harmonic for Stokes' equations without a body force,[17] the general solution for $P_n(r)$ has to be a linear combination of K_0 and I_0. Some experimentation with the second equation of (7.26) then gives Z_n, and the third equation of the group then gives R_n, so that

$$P_n = \frac{2n\pi}{h} \{ A\,K_0(r_n) + B\,I_0(r_n) \}$$

$$R_n = Ar_n\,K_0(r_n) + Br_n\,I_0(r_n) + C\,K_1(r_n) + D\,I_1(r_n)$$

$$Z_n = A\{ r_n\,K_1(r_n) - 2\,K_0(r_n) \} - B\{ r_n\,I_1(r_n) + 2\,I_0(r_n) \} + C\,K_0(r_n) - D\,I_0(r_n)$$

$$\tag{7.27}$$

where $r_n = n\pi r/h$. The values of A, B, C, and D necessary to satisfy the boundary conditions given by (7.24) and (7.25) are lengthy and are not reproduced here, but are easily obtained with standard numerical

---

[17] Take the divergence of Equation (2.24) with the nonlinear and body-force terms absent.

algebra packages.   The reader might, however, like to consider the simpler special case where the weld pool is so large that the solid/liquid boundary can be considered to be arbitrarily far away; then

$$A = -\frac{w_0 \gamma h^2}{\pi^2 a^2 \mu_E} \frac{1}{n^2 K_1\left(\dfrac{n\pi a}{h}\right)}, \quad C = \frac{w_0 \gamma h}{\pi a \mu_E} \frac{K_0\left(\dfrac{n\pi a}{h}\right)}{\left[n K_1\left(\dfrac{n\pi a}{h}\right)\right]^2}, \quad (7.28)$$

$$B = 0, \quad D = 0.$$

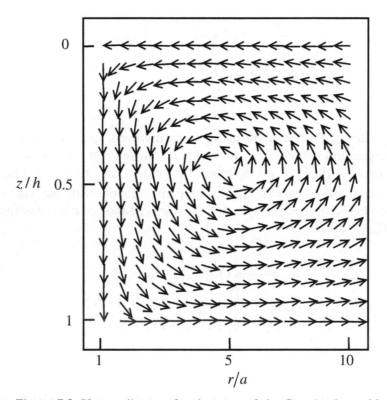

Figure 7.9. Vector diagram for that part of the flow in the weld pool caused by downflow at the keyhole wall. The case shown corresponds to $a/h = 0.1$ and a/b small. The vectors are all normalized to the same length and the arrows show the direction of flow; the keyhole wall is on the left at $r/a = 1$.

The presence of the boundary represented by the interface between the solid and liquid phases of the material of the workpiece provides an important constraint on the possible motion of the fluid, especially ahead of the laser. To ignore it reduces the dramatic effect of the circulation implied by (7.28). It represents a downward flow near to the keyhole with a return upward flow far away. If the ratio of a to h is small, the velocities near the keyhole wall are very high compared to the return velocities. The effect, however, is present but is smaller than it would be when proper account is taken of the presence of the boundary at $r = b$.

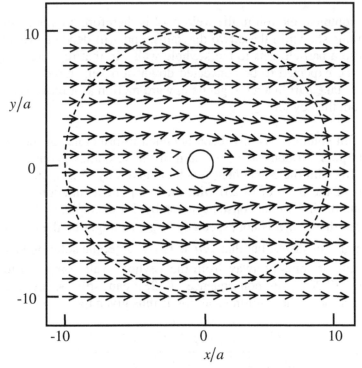

Figure 7.10. Vector diagram for that part of the flow in the weld pool caused by flow around the keyhole. The case shown corresponds to $b/a = 10$. The lengths of the vectors are all normalized to the welding speed. The solid circle is the keyhole wall and the broken circle is the liquid/solid interface.

To illustrate the presence of the effect, Figure 7.9 shows a vector diagram for that part of the flow in the weld pool caused by downflow at the keyhole wall. The value of a/h is 0.1 and a/b small. The vectors are all normalized to the same length, and the arrows show the direction of flow with the keyhole wall on the left.

For the purposes of comparison, Figure 7.10 shows a vector diagram for that part of the flow in a plane perpendicular to the laser axis. The lengths of the vectors are normalized relative to the welding speed, and the case shown corresponds to $b/a=10$. It will be seen that the motion is almost rectilinear until quite close to the keyhole wall. The velocity vector is tangential to the wall but not actually zero except at the front and rear stagnation points.

The complete motion in the weld pool is therefore a combination of the motions shown in these two diagrams, and will depend strongly on how close a particle gets to the keyhole wall. In that region it may, in general, be expected to be convected downward very rapidly and then emerge to take part in a more gentle upward motion as it moves past and away from the keyhole.

The full solution has now been obtained in principle for the flow in the weld pool. It consists of the sum of the velocities given by (7.13) and (7.18) with the component functions given by (7.22), and the sum of terms of the form of (7.23) with the component functions given by (7.27) and the individual constants given by, for example, (7.28). It will be noticed that if h/a is not especially large, as might be the case in the welding of thin sheets, but that b/a is fairly large, the solution given by (7.28) falls off rapidly away from the keyhole wall compared to that given by (7.20). In consequence, (7.28) can still be used as an approximation. The resulting solution is thus given by

$$\frac{u_r}{U} \equiv u_r' = \frac{2\ln(r'/a')(a'^4 + b'^4) - b'^2(r'^2 - a'^4/r'^2)}{2\ln(b'/a')(a'^4 + b'^4) - (b'^4 - a'^4)}\cos\theta$$

$$-U'\sum_{n=1}^{\infty}\left[\frac{r_n\,K_0(r_n)}{a_n\,K_1(a_n)} - \frac{K_0(a_n)K_1(r_n)}{K_1^2(a_n)}\right]\frac{\cos z_n}{n}$$

$$\frac{u_\theta}{U} \equiv u_\theta' = -\frac{2[\ln(r'/a')+1](a'^4 + b'^4) - b'^2(3r'^2 + a'^4/r'^2)}{2\ln(b'/a')(a'^4 + b'^4) - (b'^4 - a'^4)}\sin\theta$$

$$\frac{u_z}{U} \equiv u_z' = -U' \sum_{n=1}^{\infty} \left[\frac{r_n \, K_1(r_n)}{a_n \, K_1(a_n)} - \frac{K_0(r_n)[a_n \, K_0(a_n) + 2 K_1(a_n)]}{a_n \, K_1^2(a_n)} \right] \frac{\sin z_n}{n}$$

$$(7.29)$$

where all lengths have been made dimensionless with respect to h, $r_n = n\pi r/h$ etc. and

$$U' = \frac{\tau_0 h}{U \pi \mu_E}.$$

To calculate particle paths it is simplest to revert to the description in Cartesian coordinates in which

$$u_x' = u_r' \cos\theta - u_\theta' \sin\theta, \quad u_y' = u_r' \sin\theta + u_\theta' \cos\theta.$$

Since the particle paths are given by

$$\frac{d\mathbf{r}}{dt} = \mathbf{u}(\mathbf{r}, t),$$

all that is necessary to find typical particle paths is to solve the system of equations for a given initial position for the particle. Some typical results are shown in Figure 7.11 on the assumption that τ_0 is positive.

The positions of the starting points are shown schematically in Figure 7.12. It will be noticed that, although they are quite close together, the particle paths can be very different and remarkably complicated for a model that is, in fact, entirely linear in the molten region. It seems scarcely surprising, therefore, that observed paths can be very intricate indeed, and show such variety, when they are simply portraying the complexities of turbulent motion.

The flows associated with these examples all have a velocity at the keyhole wall that is toward the central plane. Condition (7.7) implies that there must be a region of reversed flow in the keyhole near the walls as a result. It is therefore natural to ask if this is in fact ever possible. To investigate the possibility it is necessary to have some kind of model for the flow in the keyhole. This, too, is not simple; the magnitude of the axial flow suggests that it is predominantly in the direction of the axis and is very vigorous indeed. It has already been shown that the viscous component is important, at least so far as the coupling with the molten region is concerned. It is more likely, however, that most of the bulk flow will be not unlike an inviscid motion and that it will have a strongly nonlinear character. The following model exploits the idea and also assumes that the compressibility of the vapor is not a critical factor, so that qualitative insights can be gained, but not quantitative agreement.

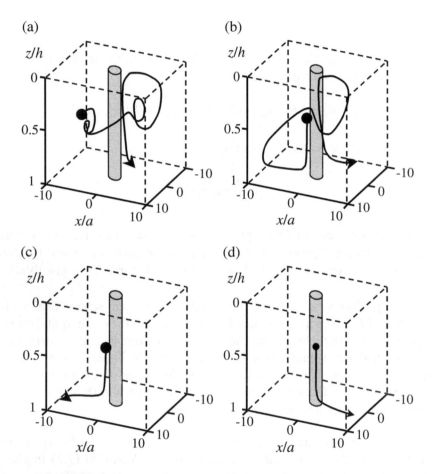

Figure 7.11. Typical particle paths in the melt pool; $h/a = b/a = 10$, $U' = 10$. All particle paths start in the plane $z/h = 0.5$, and the co-ordinates $(x/a, y/a)$ are, respectively, (a) $(-7, 0.1)$ (b) $(-1.22, 0.1)$ (c) $(-1.1, 0.01)$ (d) $(0, 1.1)$. The positions of these points are shown schematically in Figure 7.12.

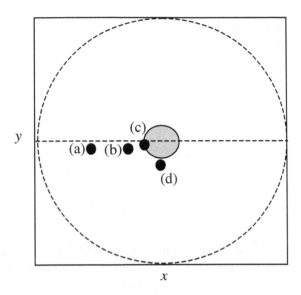

Figure 7.12. Schematic diagram showing the positions of the starting points of the particle paths shown in (a) to (d) of Figure 7.11 in the plane $z/h = 0.5$ The view is along the axis of the keyhole from above.

The mass conservation condition at the boundary is fundamental, so it, together with the definition implied in Equation (7.12), shows that in the vapor

$$u_r^V = -u_0 \sqrt{\frac{\rho_L}{\rho_V}} \text{ at } r = a. \tag{7.30}$$

Here, u_r^V is the radial component of velocity in the keyhole at its boundary. Superscript or subscript V will be used to indicate the vapor region as necessary, with L used for the liquid region.   On the assumption that u_0 is effectively constant since the keyhole radius varies little with z, this means that the radial velocity is independent of position in the keyhole parallel to the laser axis.

There is no reason to suppose that velocities in the keyhole are small, so that a nonlinear model is needed in it. As already suggested, suppose that the viscous terms are ignored, conditions are assumed to be steady, the density is taken as constant, and the velocity is assumed to have axial and radial components only. In that case, the two velocity components and the pressure satisfy the equations[18]

$$u_r^V \frac{\partial u_r^V}{\partial r} + u_z^V \frac{\partial u_r^V}{\partial z} = -\frac{1}{\rho_V} \frac{\partial p_V}{\partial r} \qquad (7.31)$$

$$u_r^V \frac{\partial u_z^V}{\partial r} + u_z^V \frac{\partial u_z^V}{\partial z} = -\frac{1}{\rho_V} \frac{\partial p_V}{\partial z} \qquad (7.32)$$

$$\frac{\partial u_r^V}{\partial r} + \frac{u_r^V}{r} + \frac{\partial u_z^V}{\partial z} = 0, \qquad (7.33)$$

which are the radial and axial components of the Euler Equations (2.17) and the equation of conservation of mass for an incompressible fluid expressed in cylindrical polar coordinates.

The conditions that need to be satisfied by this set of equations are, in principle at least,

$$p_V = p_0 \text{ at } z = 0 \text{ and } z = 2h \text{ for } 0 \le r \le a, \qquad (7.34)$$

$$u_r^V = -u_0 \sqrt{\frac{\rho_L}{\rho_V}} \text{ at } r = a \text{ for } 0 \le z \le 2h, \qquad (7.35)$$

and all quantities are bounded in $0 \le r \le a$.

The coordinate system is as before and is shown in Figure 7.8.

Equations (7.31) to (7.33) however have a solution in which u_r^V is independent of z, and u_z^V is linear in z. Although this solution does not exactly satisfy the boundary conditions (7.34) and (7.35), it is close enough to be very informative. It is simple to verify that it is given by

$$u_r^V = -u_0 \sqrt{\frac{\rho_L}{\rho_V}} \frac{af(\eta)}{r}, \quad u_z^V = 2u_0 \sqrt{\frac{\rho_L}{\rho_V}} \frac{z-A}{a} f'(\eta),$$

---

[18] Batchelor, 1967, p. 602; Landau and Lifshitz, 1959b, p.51.

$$p_V = B - \frac{u_0^2 \rho_L}{2}\left\{\frac{f^2}{\eta} + 4C^2\frac{(z-A)^2}{a^2}\right\}$$

and

$$\left(\frac{df}{d\eta}\right)^2 - f\left(\frac{d^2 f}{d\eta^2}\right) = C^2 \tag{7.36}$$

where

$$\eta = \frac{r^2}{a^2}.$$

The boundary conditions are

$$f/r \to 0 \quad \text{as} \quad r \to 0 \text{ and } f(1) = 1.$$

The two boundary conditions given by (7.34) are clearly not satisfied exactly, but it is possible to minimize the error at each end. Before that is done, however, it is necessary to study the solutions of Equation (7.36). If the boundary conditions are applied these are

$$f = \begin{cases} \dfrac{\sinh c\eta}{\sinh c} & \text{when } C<1 \text{ and } \quad C = \dfrac{c}{\sinh c} \\[2mm] \eta & \text{when } C=1 \\[2mm] \dfrac{\sin c\eta}{\sin c} & \text{when } C>1 \text{ and } \quad C = \dfrac{c}{|\sin c|}. \end{cases} \tag{7.37}$$

It is now possible to investigate the best possible choice of the parameters A, B, and C to approximate the pressure condition (7.34) at the two ends of the keyhole. On the grounds of symmetry, take $A = h$ so that $p_V(r,0) = p_V(r,2h)$. It is also essential that the pressure at the point that the plume emerges from the keyhole should be atmospheric at the edge of the keyhole, otherwise there would be a sudden change of radius of the plume at the exit. Consequently, it is necessary to have

$$B = p_0 + \tfrac{1}{2}u_0^2\rho_L\left(1 + 4C^2\frac{h^2}{a^2}\right)$$

The remaining constant c, and hence C, has to be chosen to give the best approximation possible to the pressure over the rest of the keyhole outlet. It turns out that there are important qualitative differences depending on the criterion chosen, indicating that the nature of the solution, and hence of the flow on the weld pool, can be quite sensitive to ambient conditions. It also has to be noticed that this choice may be affected by implication in the satisfying of the interface conditions at the keyhole wall when the full viscous version of the problem is studied.

Perhaps the most obvious choice is to minimize the mean square deviation at each end across the keyhole, i.e., choose c to minimize

$$2\pi \int_{r=0}^{a} (p_V - p_0)^2 \, r\,dr = \tfrac{1}{4}\pi a^2 u_0^4 \rho_L^2 \int_{\eta=0}^{1} \left\{ 1 - \frac{f^2}{\eta} \right\}^2 d\eta \equiv \tfrac{1}{4}\pi a^2 u_0^4 \rho_L^2 I \,.$$

It is now possible to draw the graph of I, which is shown in terms of the parameter c in Figure 7.13.

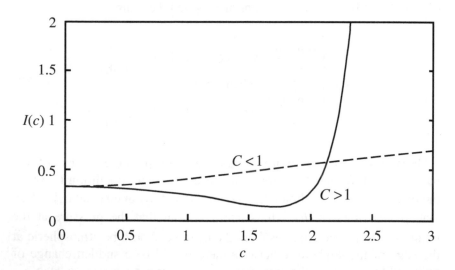

Figure 7.13. I as a function of c. See Equation (7.37) for the connection between c and C in the two cases $C>1$ and $C<1$. The value of I when $C=1$ is $\tfrac{1}{3}$ and corresponds to $c=0$.

It will be seen from this graph that there is a stationary value of I in the region $C > 1$ at $c_1 = 1.692$, with $I(c_1) = 0.13802$. The corresponding stream lines are shown in Figure 7.14. The magnitude of the reversed velocity is quite small compared to the exit velocity of the plume, but is still relatively high, corresponding to fairly high tangential velocities in the liquid region of the weld pool. There are other minima in the range $C > 1$ (not shown), and indeed for a given value of C there may be many solutions. The case when $C < 1$ increases monotonically with c to an asymptotic value of 1 and so does not provide a satisfactory solution to the problem; in addition, these correspond to solutions in which the axial velocity is predicted to be greatest at the keyhole wall. This is the region where friction effects with the surrounding liquid region might be expected to be greatest, so it is to be expected that the axial velocity is a minimum there. All those corresponding to $c \geq \frac{1}{2}\pi$ include regions of reversed axial flow in the keyhole.

The value of c_1 found by the above argument is one such solution, so it can be seen that reversed flow near the keyhole wall is indeed possible. Whether or not it occurs is likely to be strongly influenced by external conditions.

The example just considered is lengthy, but it illustrates the way in which a worthwhile model of a complicated set of interactions may be built up from simpler components. Such a model can then be used to test ideas about mechanisms that are not properly understood, or to illustrate the way in which alternative mechanisms from the accepted ones can describe observed phenomena. It is clear from the model that the result is sensitive to extraneous circumstances. The strength of the flow predicted by the model, irrespective of the direction, shows that fluid motion induced by the motion of the vapor in the keyhole must be considered very seriously. The situation is clearly complex and needs detailed investigation. Mathematical modeling in this instance does not solve any problem, but it identifies features that require further investigation.

This particular example is important to thermal modeling, although that aspect has not been dwelt on at length. The mechanism investigated, one of several, drives strong liquid motion in the direction parallel to the keyhole axis. It results in strong mixing that manifests itself in a way that is not unlike an increase in the conductivity of the liquid metal. The idea of an eddy viscosity was useful in the model, and the same idea can be applied to the conductivity. An increase in the value of the conductivity, by the introduction of an eddy conductivity, will increase theoretical estimates of the size of the weld pool. In the present example, the consequence is a larger value for b, the half-width of the melt pool, than would be expected from the value of the molecular conductivity. As a result, it can be seen that the thermal and fluid flow aspects of the problem are indeed linked.

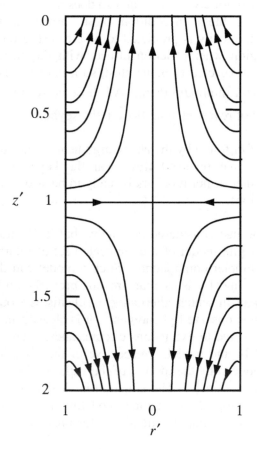

Figure 7.14. Example of stream lines of the flow in an open cylindrical keyhole in dimensionless units.

The equations of fluid dynamics and the full form of the equation of conductivity are nonlinear, and the required solutions may well be turbulent.   For accurate agreement they will have to be solved numerically, as a rule. The general considerations relating to boundary and interface conditions on these equations referred to in Section 7.1 are important, and it is advisable to read more advanced literature on fluid dynamics in depth.[19]   At a boundary between phases, the important points to remember in connection with the interface conditions are that, firstly, mass is conserved, a condition that is easily overlooked, fundamental as it is. Secondly, momentum is conserved in the absence of surface forces; it will have to be modified appropriately in the presence of such features as pressure, surface tension, or viscous forces. Thirdly, internal energy is conserved, making due allowance for latent heat effects. To these must be added appropriate extra conditions such as the isothermal condition at a melting/freezing boundary or a boiling/condensing one. Some modification to these may be necessary, depending on circumstances. Lastly, there will be conditions on the fluid motion such as the no-slip condition.

The location of the boundary is usually a part of the solution of the problem, so more conditions are to be expected than would be the case if the position of the interface were known in advance.

7.3. LASER HEATING OF THE VAPOR IN THE KEYHOLE

Energy from the laser light not only vaporizes the material in the keyhole wall, but it can also ionize the resulting vapor. As a result, a substantial interchange of energy is possible under some circumstances, indirectly from the laser beam to the molten material of the workpiece, and independently of the process of Fresnel absorption at the keyhole wall. Electrons are released and the vapor forms either a partially ionized vapor or a fully ionized plasma. The various volume elements of the plasma (if they are chosen to be large enough relative to the Debye length) tend to be electrically neutral overall. In the presence of electrons in the ionized vapor, laser light can be absorbed by a process known as inverse bremsstrahlung. Suppose that the absorption cross-section of the free electrons in a plasma is σ (m^2), then the rate of energy absorption per unit volume q (W m^{-3}) is given by

$$q = 4\pi U_{mean} \sigma n_e. \qquad (7.38)$$

---

[19] See, for example, Batchelor, 1967; Lamb, 1932; Landau and Lifshitz, 1959b.

Here, U_{mean} (W m^{-2}) is the mean intensity of radiation in all 4π directions,[20] and n_e (m^{-3}) is the number density of electrons. The absorption cross-section is given by[21]

$$\sigma = \frac{n_i Z^2 e^6 \{1 - \exp(-\hbar\omega/kT)\}}{6\mu_r \,\epsilon_0^3 \, c\hbar\omega^3 m_e^2} \sqrt{\frac{m_e}{6\pi kT}} \, \bar{g} \,. \qquad (7.39)$$

In this formula, Z is the average charge of the positive ions in units of proton charge, n_i is the number density of ions, c is the speed of light, e is the magnitude of the charge of an electron, $\hbar$ is Planck's constant divided by 2π, k is Boltzman's constant, T is the absolute temperature of the plasma (assumed to be in thermodynamic equilibrium), μ_r is the real refractive index (i.e., c times the real part of the wavenumber divided by ω), ϵ_0 is the permittivity of free space, m_e is the mass of an electron, and $\bar{g}$ is the average Gaunt factor. There are a number of formulae available for the Gaunt factor, which depends rather weakly on the frequency of the laser and the temperature in the appropriate range. Formulae are available for the classical and quantum mechanical factors,[22] and further information in some situations not covered by these is also available.[23] All these estimates suggest that a value of the order of unity is a reasonable estimate here with only a weak temperature dependence.

Consequently,

$$\sigma_0 = \sigma \left/ \frac{n_i Z^2 e^6 \bar{g}}{6\mu_r \,\epsilon_0^3 \, c\hbar\omega^3 m_e^2} \sqrt{\frac{m_e}{6\pi\hbar\omega}} \right. = \frac{1 - \exp(-1/T')}{\sqrt{T'}} \qquad (7.40)$$

where

$$T' = kT/\hbar\omega \,.$$

---

[20] Ishimaru, 1978, ch.7.
[21] Hughes, 1975, p.44.
[22] Hughes, 1975, p.41 and p.42, respectively.
[23] Berger, 1956.

Figure 7.15 shows σ_0 as a function of T'. It tends to zero at infinity with an asymptotic form given by

$$\sigma_0 \sim \frac{1}{T'^{3/2}}. \tag{7.41}$$

For a CO_2 laser, and at a temperature of 10^4 K, the value of T' is 7.4, suggesting that the asymptotic form might be adequate for modeling purposes.

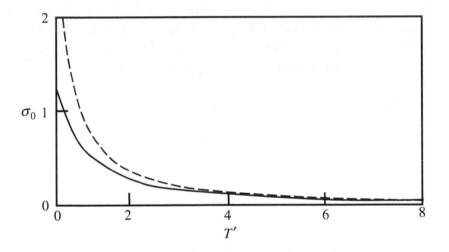

Figure 7.15. σ_0 as a function of T'. The asymptotic form is shown by the broken line.

It must not be forgotten that σ_0 does not show all the features of the dependence of σ on temperature, since it also depends on the number density of ions, n_i, which is itself dependent on T. From (7.38) it is clear that the power absorbed by the vapor also depends on the number density of electron, n_e. In any model of thermal processes in the vapor that sets out to calculate the temperature, it is necessary to relate these to the temperature. Again, on the assumption of local thermodynamic equilibrium, such a relation is given by Saha's equation,[24]

---

[24] Vicenti and Kruger, 1965; Eliezer, et al., 1986, ch. 7.

$$\frac{n_e n_i}{n_n} = \frac{K_p(T)}{kT}$$

(7.42)

in which approximately

$$K_p(T) = 2kT \frac{g_0^i}{g_0^n} \left(\frac{m_e kT}{2\pi\hbar^2}\right)^{\frac{3}{2}} \exp\left(-\frac{I}{kT}\right).$$

(7.43)

In this equation I is the ionization energy of the atoms equal to 2.524×10^{-18} J for argon, for example, and 1.26×10^{-18} J for iron; n_n is the number density of neutral ions; and g_0^i and g_0^n are the ion and neutral degeneracy factors of the groundstate.[25] The degeneracy factor for monatomic gases is given by $g_0 = (2L+1)(2S+1)$, where L and S are the orbital angular momentum and spin of the atom or ion.

Equation (7.42) can be rewritten as

$$\frac{n_e n_i}{n_n} = n_0 N(T'')$$

(7.44)

where

$$n_0 = 2 \frac{g_0^i}{g_0^n} \left(\frac{m_e I}{2\pi\hbar^2}\right)^{\frac{3}{2}}, \quad T'' = \frac{kT}{I} \quad \text{and} \quad N(T'') = T''^{\frac{3}{2}} \exp\left(-\frac{1}{T''}\right).$$

(7.45)

Figure 7.16 shows the graph of $N(T'')$; asymptotically, it is like $T''^{3/2}$.

The following relations can then be used to find all three number densities.

$$\rho_{charge} = e(Zn_i - n_e),$$

(7.46)

$$\rho = m_i n_i + m_n n_n + m_e n_e,$$

(7.47)

$$p = (n_i + n_n + n_e)kT.$$

(7.48)

---

[25] Moore, 1949.

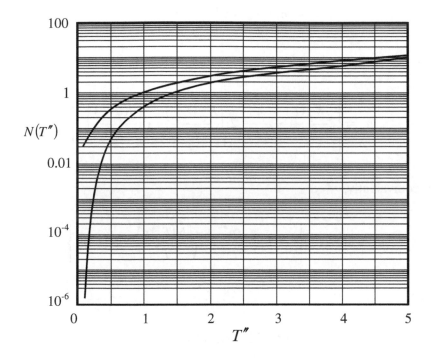

Figure 7.16. The graph of $N(T'')$ (lower curve) compared with the asymptotic form $T''^{3/2}$ (upper curve) shown on a logarithmic scale.

Here, ρ_{charge} is the electrical charge density of the vapor, p and ρ are its pressure and mass density, respectively, and m_i, m_n, m_e are the masses of an ion, a neutron, and an electron. It is reasonable to assume electrical neutrality so that $\rho_{\text{charge}} = 0$, giving

$$Zn_i = n_e,$$

and that the ions are singly ionized so that $Z = 1$. Furthermore, the approximations $m_e \ll m_i = m_n$ are worthwhile. These approximations lead to the relations

$$n_i = n_e = \tfrac{1}{2}\left\{\frac{p}{kT} - n_n\right\} \tag{7.49}$$

and

$$n_n = \left\{2\frac{K_p(T)}{kT} + \frac{p}{kT} - 2\sqrt{\frac{K_p(T)}{kT}\frac{p}{kT} + \left[\frac{K_p(T)}{kT}\right]^2}\right\} \tag{7.50}$$

or, equivalently,

$$n_n = n_0 N(T'')\left\{2N(T'') + \frac{p''}{T''} - 2\sqrt{\frac{N(T'')p''}{T''} + N(T'')^2}\right\}$$

where

$$T'' = \frac{kT}{I} \quad \text{and} \quad p'' = \frac{p}{n_0 I}.$$

The negative square root in (7.50) is essential to ensure that n_i and n_e given by (7.49) are nonnegative. A consequence of these two equations is that the mass density is given by

$$\frac{\rho}{m_n} = \frac{K_p(T)}{kT} + \frac{p}{kT} - \sqrt{\frac{K_p(T)}{kT}\frac{p}{kT} + \left[\frac{K_p(T)}{kT}\right]^2}. \tag{7.51}$$

This equation relates the pressure, temperature, and density in the plasma. When T is sufficiently small, the terms in $K_p(T)$ are negligible and the appropriate form of the perfect gas law is recovered.

From (7.38) and (7.40) it follows that

$$q = 4\pi U_{mean} Dn_e n_i \sigma_0\left(\frac{kT}{\hbar\omega}\right)$$

with

$$D = \frac{Z^2 e^6 \overline{g}(T)}{6\mu_r \in_0^3 c\hbar\omega^3 m_e^2} \sqrt{\frac{m_e}{6\pi\hbar\omega}},$$

so from (7.42) and (7.50), and taking $Z = 1$,

$$q = 4\pi U_{mean} D\sigma_0 \left(\frac{kT}{\hbar\omega}\right) \frac{K_p(T)}{kT} \times$$

$$\times \left\{ 2\frac{K_p(T)}{kT} + \frac{p}{kT} - 2\sqrt{\frac{K_p(T)}{kT}\frac{p}{kT} + \left[\frac{K_p(T)}{kT}\right]^2} \right\}, \qquad (7.52)$$

an equation that can be rewritten as

$$q_0(T'', p'') =$$

$$\frac{q}{4\pi U_{mean} D(T) n_0^2 \sigma_0(T')} = N(T'')\left\{ 2N(T'') + \frac{p''}{T''} - 2\sqrt{\frac{N(T'')p''}{T''} + N(T'')^2} \right\}. \qquad (7.53)$$

Figure 7.17 shows $q_0/p''^{3/2}$ as a function of T'' for a number of different values of p''. With the value for I given above for iron, however, T'' only has the value 0.11 at a temperature of 10^4 K, and at atmospheric pressure the value p'' is about 6×10^{-7}; the value of n_0 is of the order of 10^{29} m^{-3}. The degree of ionization tends to be low since with these values $N \approx 4.1 \times 10^{-6}$ and $p''/T'' \approx 5.5 \times 10^{-6}$. These conditions are met in the steeply rising initial part of the curve at atmospheric pressure, although there is a substantial degree of pressure dependence. Figure 7.18, therefore, shows this region in more detail; q_0/p'' is plotted for the case of atmospheric pressure, half that, and double it. Notice the very sudden rise in value from zero.

An additional problem that has to be faced in any attempt to model the temperature in the vapor is that the equation of heat conduction requires a formula for its thermal conductivity, and this itself is strongly temperature dependent. For example, in a fully ionized gas the effective conductivity[26] is approximately proportional to $T^{5/2}$. The vapor in the keyhole, however, is unlikely to be fully ionized, and indeed the degree of ionization may be quite low.

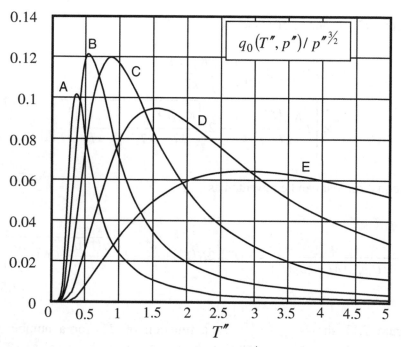

Figure 7.17. The graph of $q_0(T'', p'')/p''^{3/2}$ for a number of different values of p''. A: $p'' = 0.01$; B: $p'' = 0.1$; C: $p'' = 1$; D: $p'' = 10$ E: $p'' = 100$.

A complete form for the effective thermal conductivity at all levels of ionization is not available, but it is known that for a slightly ionized gas, the effective conductivity has the form

$$B\sqrt{T}$$

where

$$B = 8.324 \times 10^2 \times m^{-1/2} \Omega^{-1} d^{-2} \text{ J m}^{-1} \text{ s}^{-1} \text{ K}^{-3/2}$$

---

[26] Spitzer, 1962, p.144.

in which m is the molecular weight of the gas, d is the molecular diameter[27] in pm, and Ω is a parameter that varies slowly with temperature. For example, at 5000 K, Ω is 0.6, and at 10^4 K, Ω is 0.5.

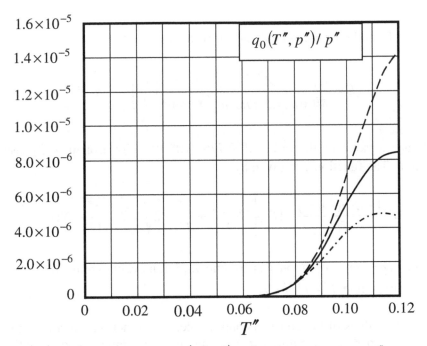

Figure 7.18. The graph of $q_0(T'', p'')/p''$ for small values of T''. The broken curve is at twice atmospheric pressure, the solid curve is at atmospheric pressure, and the lowest curve is at half atmospheric pressure.

Suppose the convective terms and the axial conduction terms in the equation of thermal conductivity are ignored and a conductivity proportional to the square root of the temperature is used. The equation then has the following structure,

$$\frac{1}{r'}\frac{d}{dr'}r'\sqrt{T''}\frac{dT''}{dr'} + PE\Sigma(T'', p'') = 0. \qquad (7.54)$$

---

[27] For iron, for example, d is about 140 pm. Emsley, 1998.

In this equation the substitution $r' = r/a$ has been used as well as the approximation

$$4\pi U_{mean} = \frac{P}{\pi a^2} .$$ (7.55)

A formula for E can be deduced from above and is

$$E = \frac{n_0^2}{\pi B} \frac{Z^2 e^6}{6\mu_r \epsilon_0^3 c\hbar\omega^3 m_e^2} \sqrt{\frac{m_e}{6\pi\hbar\omega}} \left(\frac{k}{I}\right)^{3/2} \quad \mathrm{W}^{-1}.$$

It has a value of the order of $8.775 \times 10^8 \ \mathrm{W}^{-1}$ for iron in CO_2 laser light. The function Σ is given by

$$\bar{g}(T)\frac{1 - \exp(-1/T')}{\sqrt{T'}} N(T'')\left\{2N(T'') + \frac{p''}{T''} - 2\sqrt{N(T'')\frac{p''}{T''} + N(T'')^2}\right\}.$$

The dependence of $\bar{g}$ on T is not known, but its value seems to be close to unity, while $T' = IT''/\hbar\omega$.

The values of the constants and the detailed functional forms cannot be regarded as accurate or reliable in the complicated situation prevailing in the keyhole,[28] so any calculations based on this equation can only be regarded as a rough guide. The temperature distribution on the axis must be finite and a suitable condition must be satisfied at the keyhole wall, $r' = 1$. Since the wall is an ablating boundary, the distribution of molecular velocities cannot be Maxwellian, with the result that there is a thin layer, known as a Knudsen layer, present on the boundary. It should be taken into account in a full description.[29] An approximate boundary condition, however, is that the temperature of the vaporizing metal should be at the boiling point, under atmospheric pressure, of the metal of the workpiece.

---

[28] The situation is far more complicated than is suggested by the account given here. There are many other aspects to the keyhole processes and the appropriate functional forms associated with them. For a thorough discussion of the problem see Gillies, 2000.

[29] Finke and Simon, 1990.

Equation (7.54) is a special case of a more general form that would apply with the underlying assumptions of the model, even if the detailed values of the numerical constants need to be modified. Even if it is necessary to use different formulae for the conductivity of the vapor[30] and for the absorption coefficient, it is possible to make some general remarks.   These concern the relationship between the inverse bremstrahlung contribution to the absorbed power in the workpiece, Q_B, and the laser power.

Suppose that these functions depend solely on temperature and operating pressure, and that the latter does not vary substantially across the keyhole at any given depth. Since

$$Q_B = -2\pi a \left[\lambda \frac{dT}{dr} \right]_{r=a} = -2\pi B \left(\frac{I}{k} \right)^{3/2} \sqrt{T''} \frac{dT''}{dr'} \bigg|_{r'=1}$$

and T'' is given by (7.54), which does not contain either U or a, it follows that

$$Q_B = F(P, p). \tag{7.56}$$

In other words, Q_B depends on the power at a given cross-section and the (mean) pressure at that section, as well as the numerous parameters of the materials involved and a substantial number of fundamental constants. It does not depend directly on the radius of the keyhole and, hence, on the spot size or the welding speed.

An example is shown in Figure 7.19. It corresponds to atmospheric pressure in the keyhole and a power of 5 kW at the given cross-section. The temperature at the keyhole wall is the boiling point of iron. The temperature on the axis necessary to maintain this state is 0.216 dimensionless units, corresponding to a temperature of 19600 K. The dimensionless value of the power at the wall is $\sqrt{T''}\, dT''/dr' = 1.366$, corresponding to a power per unit depth of 3.9 kW cm^{-1}

A remarkable feature of Equation (7.55) is that, for a given power, there are three solutions if the power is sufficiently high. One solution is that in which the temperature is almost uniformly equal to the

---

[30] The analysis here, for example, uses a formula appropriate to low degrees of ionization.   In some circumstances it might be more appropriate to use a generalized formula that could incorporate high ionization effects as well.

temperature at the keyhole wall. With the values used for the case of iron, there is no other solution for a power P of less than about 0.82 kW. For greater values there is an upper solution with temperatures rising rapidly to the range 11-20 kK, and a middle branch with temperatures in the range 7-11 kK. In the first case the power absorbed, Q_B, rises with the power, but in the second case it falls very slowly. It is interesting that experimental evidence is divided on the question of keyhole temperatures. It is a very difficult quantity to measure. Most measurements favor low values, around 5000K, but some workers have observed higher values, up to 16-18 kK. Most theoretical work has tended to predict values in the highest range, and it is interesting that the analysis provided here does not exclude either possibility. A stability analysis in which the time dependence of the conduction equation is taken into account would, however, indicate which of these solutions are stable, and therefore more likely to occur. At low powers when no other solution exists, the near-isothermal solution is undoubtedly stable, so it is probable that the intermediate temperature-range solution is unstable and the high temperature-range solution is stable.

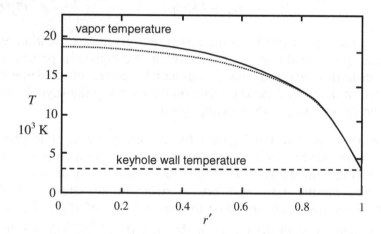

Figure 7.19. $T(r')$ for iron at atmospheric pressure and a power of 5 kW. The dotted line includes a correction for a high degree of ionization.

Figure 7.20 shows the temperature on the axis as a function of power P. Similarly, Figure 7.21 shows the power absorbed, Q_B, as a function of P. Each figure shows both possible solutions. The upper branch in Figure 7.21 is very well described by the empirical fit

$$Q_B = Q_{\mathfrak{L}} + \sqrt{\mathfrak{L}(P - P_{\mathfrak{L}})} \tag{7.57}$$

where $Q_{\mathfrak{L}} \approx 1\,\text{kW cm}^{-1}$, $P_{\mathfrak{L}} \approx 1\,\text{kW}$, and $\mathfrak{L} \approx 1.8\,\text{kW cm}^{-2}$. $P_{\mathfrak{L}}$ and $Q_{\mathfrak{L}}$ provide a threshold below which there is effectively no transfer of power from the laser to the workpiece by this mechanism. Equation (7.57) is thus a particular case of Equation (7.56) and a generalization of (6.18). $\mathfrak{L}$ in Equation (7.57) is therefore a generalization of the Linking Intensity concept introduced in (6.18).

The lower branch of the curve gives a value for Q_B that does not differ very much from $200\ \text{W cm}^{-1}$ over most of the range shown, but can be described reasonably well for $1 < P < 20\,\text{kW}$ by a similar formula to (7.57) in which

$$Q_B = Q_{\mathfrak{L}} - \sqrt{\mathfrak{L}(P - P_{\mathfrak{L}})}$$

with $Q_{\mathfrak{L}} \approx 300\ \text{W cm}^{-1}$, $P_{\mathfrak{L}} \approx 1\,\text{kW}$ and $2\ \text{W cm}^{-2}$.

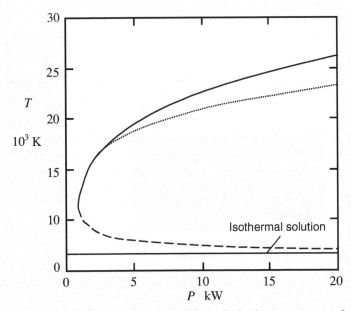

Figure 7.20. The temperature on the axis for iron vapor as a function of the power at the given cross-section of the keyhole. The dotted line includes a correction for a high degree of ionization.

The results shown here are all calculated using the formula

$$\lambda = B\sqrt{T}$$

for the thermal conductivity. It corresponds to a low degree of ionization, and is therefore appropriate near the wall of the keyhole.

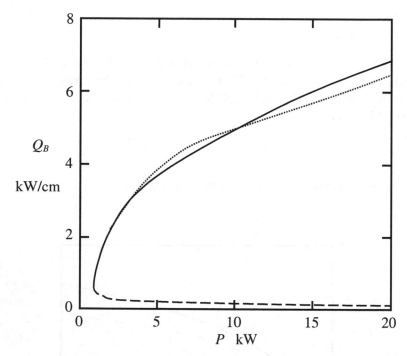

Figure 7.21. Power absorbed per unit depth at the keyhole wall for iron vapor as a function of power at the given cross-section of the keyhole. The empirically fitted relation (7.57) for the upper branch of the curve is not shown, as it is indistinguishable on the scale of the graph from the theoretically derived curve. The dotted line includes a correction for a high degree of ionization.

At high degrees of ionization the formula

$$\lambda = AT^{5/2}$$

is more appropriate. In this formula[31]

---

[31] Spitzer and Härm, 1953.

$$A = \frac{640\,\epsilon_0^2}{e^4\,\ln\Lambda}\sqrt{\frac{2\pi k^7}{m_e}}\,\frac{\epsilon\delta_T}{Z} \quad \text{with } \Lambda = \frac{3\pi}{e^3}\sqrt{\frac{(2\epsilon_0\,kT)^3}{n_e}}\,\frac{1}{Z}\sqrt{\frac{2}{1+Z}}\,.$$

For[32] $Z = 1$ and $T = 2\times10^4$ K, $\epsilon\delta_T/Z = 0.0943$, $\ln\Lambda = 4.06$ and $A = 4.54\times10^{-11}$ J m^{-1} s^{-1} K$^{-7/2}$. The two formulae for the thermal conductivity give the same value at

$$T = \sqrt{B/A} \approx 15000 \text{ K}.$$

It will be seen that this temperature is reached on the axis in the power range under consideration. It is known that the transition from one formula to the other is extremely complicated. However, if the solutions are recalculated using the low ionization formula for temperatures below 15 kK and the high ionization formula for temperatures above it, the results are not very different. Figures 7.19 to 7.21 show the result of recalculation with

$$\lambda(T) = AT^{5/2}H\!\left(T - \sqrt{\frac{B}{A}}\right) + B\sqrt{T}H\!\left(\sqrt{\frac{B}{A}} - T\right)$$

where H is the Heaviside step function. Only the upper branches of the curves in Figures 7.20 and 7.21 are affected, and then only for values of P greater than about 2 kW. The maximum temperatures are slightly lower by an amount that increases gradually to about 10% at 20 kW. The value of Q_B is slightly less at the top of the range but slightly greater between about 4 kW and 8 kW. The differences cannot be regarded as significant in view of all the uncertainty in the factors and formulae used in constructing the equation.

The general result given by (7.56) is of interest in itself. The specific results depend on fairly substantial assumptions about the constants appropriate to the specific materials involved. In general, however, the relationship has the properties that

- it depends on the properties of the constituents of the workpiece, the shielding gas, or any surrounding atmosphere that might become entrained in the keyhole;

---

[32] Numerical values taken from Dowden et al., 1989.

- it depends on the pressure in the keyhole, which will normally be close to the ambient pressure;[33]
- it depends on the frequency of the laser employed;
- it does not depend on axial convection in the keyhole, which is assumed to be entirely dominated by radial conduction;[34]
- it does not depend on the radius of the keyhole or the welding speed.

The form of F in (7.56), the linking intensity in its original form given by (6.18) or its modified form given in (7.57), must all be regarded as quantities that have to be determined empirically at present. The procedure described here can provide values in principal. There is, however, a very complex mixture of materials and states of ionization in any real keyhole. Uncertainties in the correct numerical values to use and in determining precisely which formulae are appropriate suggest that any such formulae are best regarded as no more than indications of orders of magnitude and general form.

FURTHER READING

The weld pool

Clucas et al., 1996; Dowden et al., 1996, 1998; Gellert et al., 1995; Kapadia et al., 1998; Klein et al., 1994, 1996; Ol'shanskii et al., 1974; Paulini and Simon, 1993; Paulini et al., 1990, 1993; Vicanek et al., 1994.

The keyhole

Beyer et al., 1995; Cram, 1985; de Groot, 1952; Dowden et al., 1991, 1993, 1994; Ducharme et al., 1993; Ferlito and Riches, 1993; Kapadia et al., 1996, 1997; Landau and Lifshitz, 1980; Liboff, 1990; Martin and Bowen, 1993; Matsunawa and Ohnawa, 2000; Melchior, 1981; Rodden et al., 2000; Sitenko and Malnev, 1995; Tix and Simon, 1993; Tix et al., 1995; Wang and Uhlenbeck, 1945; Whipple and Chalmers, 1944.

---

[33] But not necessarily in the case of laser welding in a vacuum, where the pressure has to be related to fluid dynamic effects.

[34] Equivalent to assuming that only a small fraction of the laser power is carried by the vapor out of the ends of the keyhole.

CHAPTER 8

THERMOELASTIC PROBLEMS

_____

A very simple example can be used to show the way in which changes in temperature can produce distortion as a result of changes in volume. The example that follows shows a number of ways in which the linear theory of thermoelasticity can be used to investigate unusual problems.

8.1 THERMAL EXPANSION

Consider the following very simple model for the melting of a solder wire on a soldering iron. The wire is fed steadily and continuously onto the surface of the soldering iron, which melts it. The molten metal drains rapidly away. If the wire is at temperature T_0 far away and T_M at the soldering iron, and the speed at which the wire is fed to the iron is U, then the temperature $T(x)$ of the wire as it approaches the iron is given by the equation of heat conduction. In this case it has the very simple form

$$\kappa \frac{d^2 T}{dx^2} = U \frac{dT}{dx},$$

(8.1)

where κ is the thermal diffusivity of the solder. If conditions are steady, figure 8.1 shows the basic configuration and the coordinate system.

Suppose the wire has a coefficient of linear expansion α_L so that a short length of wire, whose temperature is ℓ at temperature T_0, has a length $\ell\{1 + \alpha_L (T - T_0)\}$ at temperature T. Then, bearing in mind that $\alpha_L = \frac{1}{3}\alpha$ where α is the volume coefficient of expansion,[1] the relation between the stress tensor, which describes the forces in the wire, and the local displacement of elements at a given point relative to their positions if the wire is unheated, is given by Equation (2.25). In this instance the

_____

[1] See p.46.

only displacement of interest is along the line of motion; suppose it is given by $\xi(x)$.

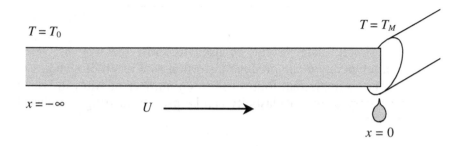

$T = T_0$ $T = T_M$

$x = -\infty$ $U \longrightarrow$

$x = 0$

Figure 8.1. The melting of solder.

The $(1,1)$ component of the stress-strain relation (2.25) shows that

$$E \frac{d\xi}{dx} = p_{11} + E\alpha_L(T - T_0) \tag{8.2}$$

where E is Young's modulus and T_0 is the temperature of the wire far from the soldering iron. If there is no stress in the wire, $p_{11} = 0$. It is then possible to find the amount X by which an element of the wire just reaching the soldering iron is ahead of the position it would occupy if there were no soldering iron present (see Figure 8.2). The solution of (8.1) for T is

$$T = T_0 + (T_M - T_0)\exp\left(\frac{Ux}{\kappa}\right). \tag{8.3}$$

Solving (8.2) shows that

$$\xi = \alpha_L(T_M - T_0)\int_{-\infty}^{x} \exp\left(\frac{Ux}{\kappa}\right) dx$$

and therefore

$$X = \frac{\alpha_L (T_M - T_0) \kappa}{U}.$$

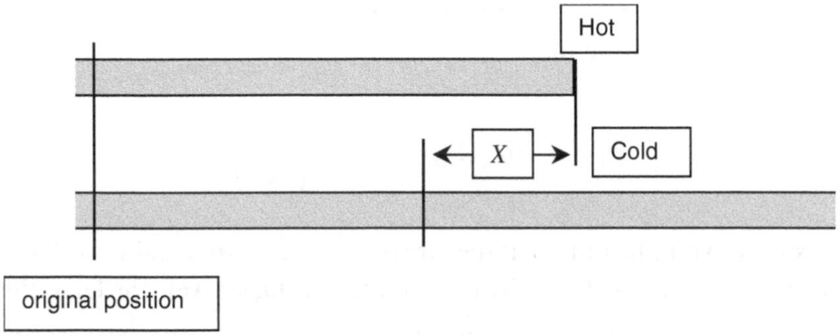

Figure 8.2. Displacement due to thermal expansion.

If the values of the constants for lead given in Appendix 1 are used as a guide, $X = \xi(0)$ is of the order of $0.2/U$ mm if U is measured in mm s^{-1}. Unsurprisingly, it is a small amount.

In this example, it is more than likely that convective cooling of the wire will have a significant effect. Allowance could be made for it by assuming a Newtonian cooling law in which the rate of removal of heat from the wire is proportional to the local excess temperature. In that case, a term of the form $c(T - T_0)$ might be added to the right-hand side of (8.1) where the value of c will depend on the diameter of the wire. In numerical terms, the only difference to the result is effectively to multiply the U by a factor $\frac{1}{2}\left\{1 + \sqrt{1 + 4\kappa c/U^2}\right\}$, resulting in an even smaller thermal extension.

Notice that the mathematical model allows any positive value for U, in principle at least. There is, therefore, an assumption that the soldering iron is capable of supplying as much heat as is necessary to melt the solder, whatever the rate of delivery. It is therefore somewhat different from typical problems of the kind in which an incident laser beam is the power source, so that the power input at $x = 0$ is also specified. That determines the feed-rate U as well.[2]

---

[2] See Chapter 5.

By way of contrast, consider a second example in which it is assumed that the wire is somehow confined so that it cannot expand along its length. The thermal conditions, however, are assumed to be the same so that the temperature in the wire is still given by Equation (8.3). The component p_{11} of the stress tensor is still given by (8.2), but this time $\xi \equiv 0$ showing that

$$p_{11} = -E\alpha_L(T_M - T_0)\exp\left(\frac{Ux}{\kappa}\right). \tag{8.4}$$

This is negative indicating that the stress is compressive and its value at $x = 0$ is $-E\alpha_L(T_M - T_0)$. With the typical values given for lead, this has a magnitude of about 1.3×10^9 Pa. Even bearing in mind that the area of cross-section of a wire of diameter 1 mm is about $10^{-6}\,\text{m}^2$ this would still represent a force of about 10^3 N. The calculation provides a warning that the forces in a heated workpiece can be very substantial and cannot necessarily be ignored. In applications such as welding they can present a problem, but in such processes as laser forming they can be turned to advantage. That is the case with the process described next, in which the existence of such forces is all-important for the success of the technique.

8.2 THE SCABBLING OF CONCRETE

A major problem in the decommissioning of nuclear power stations is to remove the large concrete shields installed around the radioactive portions of the plant. A process known as *scabbling* is used. The word "scabble"[3] is a stonemason's term meaning to work without finishing so that a block has the same appearance as stone does before it has left the quarry. By extension, it is now used to mean producing a rough surface on a piece of worked stone or concrete, and in the present context is used to describe the removal of the surface layer. In the decommissioning of a nuclear power plant[4] all that is necessary is to remove the radioactive surface layer. This is an expensive and potentially hazardous operation, but it only requires the removal of a few centimeters of material. Afterwards, the remaining bulk of the

---

[3] Or "scapple."
[4] Li et al., 1994.

material, a meter or more thick, can be removed by relatively inexpensive traditional techniques. At the time of writing, the technique for removal of the surface layer is purely mechanical and involves bulky and expensive machinery to be present in the radioactive area. That, itself, becomes radioactive, and its disposal then also becomes a problem. It has been shown, however, that a laser can be used to produce the same effect[5] (see Figure 8.3 for an example).

Figure 8.3. An example of the laser scabbling of concrete.

The virtue of a laser system is that the laser power can be generated remotely and delivered to the operational area using relatively inexpensive components in the immediate vicinity of the radioactive material. There is a much reduced problem of disposal of valuable equipment that has become radioactive as a result. The process has been evaluated experimentally in the laboratory and it was found that it is not wavelength dependent, both CO_2 and Nd-YAG lasers proving successful in the surface removal of concrete. An alternative technique based on the use of lasers, with application to both lasers and metals, is to use higher intensities with the intention of ablating the surface material.[6] The removal regime is substantially different and results in

---

[5] Johnston et al., 1999.
[6] Savina et al., 1999.

particles being melted in the process of removal and the neat cement mix being ablated. The scabbling technique uses a fairly low intensity beam applied for a short time over a large area in the form of a laser beam moving relative to the workpiece. Material is removed as solid flakes of concrete; these are ejected from the surface explosively. If the surface of the material becomes hot enough to melt, the efficiency of the process is dramatically reduced, so there is a fundamental difference between this technique and the ablation technique. Experiments also showed that if the workpiece were allowed to cool and the treatment repeated, the process did not result in significantly more material being removed. For all these reasons it is important to understand the mechanisms involved and to find what process parameters will optimize the depth of removal.

Although the way in which the scabbling takes place is not entirely certain, one way to proceed in understanding it is to propose a mechanism and study its consequences theoretically. If the results are clearly at variance with experiment, then the mechanism must be rejected. Agreement with experimental results, initially qualitative and hopefully at a later stage quantitative as well, does not necessarily prove that the mechanism is correct, but at least lends credence to it and shows that it is worth further consideration. As an illustration of this approach, consider the suggestion that the mechanism is based on the assumption that the laser beam causes the surface layer of the concrete to expand as it is heated. Only a very thin layer is directly caused to expand by the heating, but it results in large internal stresses in the concrete. Not all such stresses are compressive; very considerable tensile stresses can occur in the region under the surface layer. Concrete is relatively weak to tensile stress in comparison with its very considerable ability to withstand compressive stresses. A consequence is that stresses can be relieved by subsurface fracture, followed by ejection of the material above the fracture. An irregularly shaped crater is then left in the surface of the workpiece.

To study the process theoretically, it is possible to construct a mathematical model on a small number of physical principles. First, one can assume that the temperature distribution induced by the laser beam incident on the surface of the concrete gives rise to thermally-induced stresses. Second, assume that the linear theory of elasticity can be used to find the stress distribution in the block, from which it is possible to deduce the regions of maximum tensile stress. One can then use an appropriate criterion for the onset of fracture to find the circumstances under which fracture might occur. From such information, one can make estimates for the likely depth of scabbling. It is important to be able to establish this depth, as it is known experimentally that, in practice, only a single pass is possible, and that the shape of the incident beam is critical.

The method described becomes particularly simple in connection with laser scabbling of concrete, but the approach can be applied much more generally. For example, the stresses induced during surface heat-treatment of metals can be studied in this way or, at a certain level of approximation, their occurrence in welding problems. In the case of the scabbling of concrete, there is only a very thin thermal layer at the surface, resulting in a simplification of the analysis.

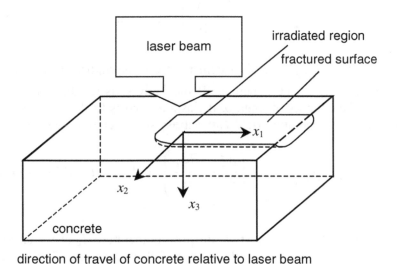

Figure 8.4. The orientation of the scanning laser beam relative to the concrete block.

If the block is scanned at a steady speed U by the laser beam, the temperature in it is given by the equation of heat conduction,

$$U\frac{\partial T}{\partial x_1} = \kappa\left(\frac{\partial^2 T}{\partial x_1^2} + \frac{\partial^2 T}{\partial x_2^2} + \frac{\partial^2 T}{\partial x_3^2}\right), \tag{8.5}$$

in which T is the temperature in the block above the ambient value, U is the velocity of the workpiece relative to the laser beam in the direction of the positive x_1-axis, κ is the thermal diffusivity of the concrete, and (x, y, z), or equivalently (x_1, x_2, x_3), are coordinates measured with the axis of the beam as the positive x_3-axis and the origin in the surface of the workpiece. See Figure 8.4.

If a body possesses a stress tensor[7] $p = \{p_{ij}\}_{i,j=1..3}$, which is necessarily symmetric, and it is subject to no body forces and the strains are small,[8] then it satisfies the equations

$$\sum_{j=1}^{3}\frac{\partial p_{ij}}{\partial x_j} = \rho\left(U\frac{\partial}{\partial x_1}\right)^2\xi_i, \quad i = 1, 2, 3. \tag{8.6}$$

Here, $\xi = (\xi_1, \xi_2, \xi_3)$ are displacements from the undisturbed state and ρ is the density of the concrete. If the body satisfies the stress-strain relation of the theory of linearized elasticity and is furthermore thermoelastic in the temperature field T, the stress-strain relations are given by

$$p_{ij} = \lambda_L\delta_{ij}\sum_{j=1}^{3}\frac{\partial\xi_j}{\partial x_j} + \mu_L\left(\frac{\partial\xi_i}{\partial x_j} + \frac{\partial\xi_j}{\partial x_i}\right) - (3\lambda_L + 2\mu_L)\alpha(T - T_0)\delta_{ij}.$$

$$\tag{8.7}$$

It is assumed that there exists a uniform temperature T_0 at which there are no stresses and no displacements. See Equation (2.36). The constants λ_L, μ_L are the Lamé constants, related to the more familiar Young's modulus E and Poisson's ratio ν by

$$E = \frac{\mu_L(3\lambda_L + 2\mu_L)}{\lambda_L + \mu_L} \text{ and } \nu = \frac{\lambda_L}{2(\lambda_L + \mu_L)};$$

---

[7] See Section 2.1 for the definition of the stress tensor and the main properties of Cartesian tensors, or e.g., Jeffreys, 1957; Hunter, 1983, 69-76. See Boley and Weiner, 1960, or Nowacki, 1986 for theory of thermoelasticity.
[8] For the definition of the strain tensor see Section 2.1.

See Equation (2.35).

Typical operating speeds for the scabbling process are of the order of 10 cm s^{-1} for U with a beam width of 2-5 cm. The Lamé constants are of the same magnitude as Young's modulus. For concrete its value is of the order of 10^{10} Pa. On the assumption that the displacement terms and the thermal terms in (8.7) are of the same order of magnitude, X, the ratio of the left-hand side of Equation (8.6) to the right-hand side is roughly

$$\frac{1}{0.025}\times10^{10}\times\frac{X}{0.025}:(0.1)^2\frac{X}{(0.25)^2}=10^{12}:1,$$

from which it is clear that the inertial terms can be safely neglected. When (8.7) is substituted into (8.6) with the neglect of these terms, the equation can be taken as

$$(\lambda_L+\mu_L)\sum_{j=1}^{3}\frac{\partial^2\xi_j}{\partial x_i\partial x_j}+\mu_L\sum_{j=1}^{3}\frac{\partial^2\xi_i}{\partial x_j^2}=\tfrac{1}{3}\alpha(3\lambda_L+2\mu_L)\frac{\partial T}{\partial x_i},\quad i=1...3$$

$$(8.8)$$

in which α is the volume coefficient of expansion.[9]

It is important to apply the appropriate boundary conditions. For the problem under consideration, in which a steady-state solution is required for $-\infty < x, y < \infty, 0 \le z < \infty$, these are

$$-\lambda_c\frac{\partial T}{\partial z}\bigg|_{z=0}=I(x,y),$$

$$\sum_{j=1}^{3}p_{ij}n_j=0 \text{ at } z=0,\qquad(8.9)$$

$$\xi\to0, T\to T_0 \text{ as } |\mathbf{r}|\to\infty.$$

---

[9] Given in terms of the linear coefficient of expansion α_L, which is often quoted, by $\alpha = 3\alpha_L$; see Equation (2.30) and the discussion of it.

In these conditions, λ_c is the thermal conductivity of the concrete and $\mathbf{n}$ is the direction of the normal. Since the surface is given by $z = 0$, then with the usual convention that the normal points out of the material, it has the components $(0,0,-1)$.

In order to solve Equation (8.8) it is necessary to have the solution of Equation (8.5) for the temperature. Suppose that there is an incident intensity distribution $I(x,y)$ on the plane surface at $z = 0$ of a semi-infinite concrete block. Some order of magnitude calculations are helpful at this point. The process can be modeled using a thin-layer approach for the thermal surface layer. In the layer, the vertical scale for variation in χ is the same as that for T; in that case the ratio of the vertical to horizontal length scales is $1/\sqrt{\text{Pe}}$, where Pe is a Péclet number based on the horizontal length scale. With a typical value for U of 10 cm s^{-1} and a horizontal length scale for the incident radiation of 5 cm, this ratio is of the order of 10^{-2} or less. It seems likely, therefore, that the contribution from χ at depth is relatively unimportant. In that case it is only necessary to find ζ at depth. The fact that the vertical length scale is so small in relation to the horizontal length scale means that the approximation discussed in Section 3.2 can be used to calculate the temperature distribution. It is therefore possible to find the value at the surface of the temperature and its derivatives.

The stress tensor $\{p_{ij}\}$ can then be found from its form (2.33) given in terms of the displacements. From Equation (8.8), bearing in mind the surface conditions (8.9), it follows that, with typical operational values, the stress components have magnitudes of the order of 2×10^6 N m^{-2}, which is of the order of the maximum tensile stress that concrete can support. Typical values are of the order of $3 - 4 \times 10^6$ N m^{-2}.

In principle, these equations can be solved by taking the Fourier transform in x and y of the surface intensity. It is simpler, however, to construct the appropriate point distribution for the displacement and so find the solution for the displacements for a specific problem in terms of a double convolution over the surface S. The surface stress condition (8.9) must be used in the process. The detailed formulae for the point functions, and the resulting convolutions giving the solution for any given incident intensity, will be found in Appendix 2.

The same approach is equally useful in other contexts, such as the study of stress distributions produced by heating a semi-infinite metal workpiece.

8.3 TWO-DIMENSIONAL MODELS

Experimental experience has shown that a relatively broad scanning beam or array of beams is far more effective than a smaller one. An incident region of perhaps a couple of centimeters from front to back and several centimeters wide has proved to be effective. Compare the intensity profiles for the circular Gaussian beam shown in Figure 3.5 with the model distribution for an array given by Equation (3.12) and shown in Figure 3.6. It is clear that the wide beam leads to a region that is correspondingly wide with little variation in temperature or intensity across it. By way of contrast, in the case of the Gaussian beam there is substantial variation. In effect, most of the region affected by the induced thermal stresses is affected in a way that may be expected to be largely independent of the lateral coordinate x_2. This leads to the possibility of simplifying the model very considerably by using a two-dimensional approximation in which the stresses and strains are regarded as independent of the x_2 component. The simplification is considerable, all the more so in that it is possible to work in a very simple manner directly in terms of the stresses without needing to solve for the displacements. Their values, in this problem at least, are largely irrelevant.

There are in fact a couple of two-dimensional approximations available in the linear theory of elasticity: the plane stress model and the plane strain model. They are appropriate to different circumstances and it is necessary to decide which is the more appropriate one for use in any given case. They both apply in the case of a steady problem in which three conditions apply. The first is that the components of the stress and strain tensors and the temperature have no dependence on the x_2 coordinates, the second is that $p_{12} = p_{32} \equiv 0$, and the third is that the body forces are derivable from a potential so that $\rho\mathbf{F} = -\nabla\Omega$ where Ω

also has no dependence on x_2. In problems considered here, however, there is no separate body force $\mathbf{F}$.

The two approximations are
- *Plane strain* in which the assumption is made that $e_{22} \equiv 0$.
- *Plane stress* in which it is required that $p_{22} \equiv 0$.

In the situation considered here, the surrounding concrete is constraining the heated region so that it is difficult for displacements to take place transversely relative to the direction of the scanning beam. Consequently, e_{22} may be expected to be zero, but not p_{22} since it is this component of stress that prevents lateral displacement. The plane strain model is therefore the one that is appropriate here.[10]

The plane strain approximation makes it possible to solve directly for the stress components in terms of the Airy stress function, χ. In the absence of any body forces, it follows from (2.41) that χ has the properties

$$p_{11} = \frac{\partial^2 \chi}{\partial z^2}, \quad p_{13} = -\frac{\partial^2 \chi}{\partial x \partial z}, \quad p_{33} = \frac{\partial^2 \chi}{\partial x^2}, \tag{8.10}$$

and from Equation (2.43) that it satisfies the equation

$$\left(\frac{\partial^2}{\partial x^2} + \frac{\partial^2}{\partial z^2} \right)^2 \chi + \frac{E\alpha}{3(1-v)} \left(\frac{\partial^2}{\partial x^2} + \frac{\partial^2}{\partial z^2} \right)(T - T_0) = 0. \tag{8.11}$$

To solve this equation it is only necessary to find a solution of

$$\left(\frac{\partial^2}{\partial x^2} + \frac{\partial^2}{\partial z^2} \right)\chi_1 + \frac{E\alpha}{3(1-v)}(T - T_0) = 0 \tag{8.12}$$

and add to it the general solution χ_2 of the biharmonic equation

---

[10] By contrast, a thin plate unconstrained on its surfaces may be assumed to have very little variation in the stresses perpendicular to its plane, but there may be nonzero strains so that the plane stress approximation might be more appropriate.

$$\left(\frac{\partial^2}{\partial x^2}+\frac{\partial^2}{\partial z^2}\right)^2 \chi_2 = 0. \tag{8.13}$$

Since

$$\frac{\partial T}{\partial x}=\frac{\kappa}{U}\left(\frac{\partial^2}{\partial x^2}+\frac{\partial^2}{\partial z^2}\right)T$$

from Equation (8.5), differentiation of (8.12) with respect to x and elimination of $\partial T/\partial x$ show that

$$\left(\frac{\partial^2}{\partial x^2}+\frac{\partial^2}{\partial z^2}\right)\left(\frac{\partial \chi_1}{\partial x}+\frac{E\alpha\kappa}{3(1-v)U}(T-T_0)\right)=0,$$

so that an appropriate solution for χ_1 is

$$\chi_1 =-\frac{E\alpha\kappa}{3(1-v)U}\int_{-\infty}^{x}(T-T_0)dx.$$

Notice that χ_1 has been given the property that it is zero far ahead of the laser beam when $T=T_0$. The surface boundary condition is one of zero normal stress so, bearing in mind the relation between χ and $\underline{\underline{p}}$ given in Equation (8.10), the conditions to be satisfied by χ_2 are

$$\frac{\partial^2 \chi_2}{\partial x\partial z}=\frac{E\alpha\kappa}{3(1-v)U}\frac{\partial T}{\partial z} \text{ and } \frac{\partial^2 \chi_2}{\partial x^2}=\frac{E\alpha\kappa}{3(1-v)U}\frac{\partial T}{\partial x} \text{ at } z=0.$$

Use has been made of the definition of χ as $\chi_1+\chi_2$ and (8.10). Because of the boundary condition

$$-\lambda_c \frac{\partial T}{\partial z}=I(x) \text{ at } z=0,$$

the fact noted above, that for concrete under typical operational conditions for scabbling the thermal layer is much thinner than the stress layer, means that the right-hand side of the second condition can be neglected compared to the right-hand side of the first. Consequently, the boundary conditions on χ_2 are

$$\frac{\partial^2 \chi_2}{\partial x \partial z} = -\frac{E\alpha\kappa}{3(1-v)U\lambda_c}I(x) \text{ and } \frac{\partial^2 \chi_2}{\partial x^2} = 0 \text{ at } z = 0,$$

$$\chi_2 \to 0 \text{ as } |x|, z \to \infty,$$

and χ_2 satisfies (8.13). In the body of the block, where the thermal effects are negligible, knowledge of the form of χ_2 is all that is needed to calculate the stress distribution.

The stress tensor $\{p_{ij}\}_{i,j=1..3}$ at any depth can now be found from the equation. Regard its components as the components of a matrix P. In order to study what happens it is necessary to note that concrete is very strong indeed to compressive stress but relatively weak to tensile stress. To identify these regions, consider the tractions on a unit disk in the material perpendicular to the direction of a unit vector **n** as its orientation is allowed to vary. The direction of **n** in which this has the largest value is the direction of maximum stress, measured positive when it is tensile and negative when it is compressive. The method of Lagrange multipliers shows very simply[11] that **n** is the eigenvector of P corresponding to the largest eigenvalue p which, for the stress to be tensile, must be positive, i.e., $P\mathbf{n} = p\mathbf{n}$ with $p > 0$.

From analysis of this kind, it is possible to predict the orders of magnitude of the surface temperatures and the internal stresses in the concrete. For concrete with limestone aggregate the tensile strength is of the order of $3 - 4 \times 10^6 \text{ N m}^{-2}$, while with basalt aggregate it is rather higher at $6 \times 10^7 \text{ N m}^{-2}$. The glazing temperature is about $1300°C$, so the intensity must be low enough, or the speed high enough, not to cause glazing. On the other hand, the intensity must be high enough, or the speed low enough, for a sufficiently high tensile stress to develop in the interior of the workpiece. These limits provide constraints on the operating parameters. The model predicts surface temperatures of about 400-800 K above ambient and internal stresses of the order of 10^7 N m^{-2}. It is thus consistent with experiments, and that suggests scabbling can occur without glazing under these conditions. It is more difficult to predict the regions in which fracture might occur. It is, however, possible to obtain exact analytical solutions of Equation (8.13) for

---

[11] For details of the argument see Section 2.1.

certain patterns of incident intensity, and these can be very helpful in assessing possible behavior of the concrete.

Make the equations dimensionless by the substitutions

$$x' = x/a, \ z' = z/a, \ \chi_2' = \chi_2 \frac{3(1-v)U\lambda_c}{E\alpha\kappa a Q_0}, \ I' = \frac{a}{Q_0} I$$

where

$$Q_0 = \int_{-\infty}^{\infty} I(x)dx$$

and a is a measure of the width of the beam. In that case the problem to be solved is to find χ_2' satisfying the conditions

$$\left(\frac{\partial^2}{\partial x'^2} + \frac{\partial^2}{\partial z'^2} \right)^2 \chi_2' = 0,$$

$$\frac{\partial^2 \chi_2'}{\partial x'\partial z'} = -I'(x') \text{ and } \frac{\partial^2 \chi_2'}{\partial x'^2} = 0 \text{ at } z' = 0,$$

$$\chi_2' \rightarrow 0 \text{ as } |x'|, z' \rightarrow \infty.$$

The biharmonic equation has a general solution[12] that can be written as

$$\chi_2' = \text{Re}\{(x' - iz')f(x' + iz') + g(x' + iz')\}$$

where $i^2 = -1$ and f and g are complex functions. For some special forms of I' it is possible to find suitable functions f and g.

Three examples are given. To derive their forms without prior knowledge is not immediate, but it is easy to verify in a matter of minutes that they have the correct properties using a suitable computer algebra package. The first example (I) is the two-dimensional version of a Top Hat profile with an incident intensity $\frac{1}{2}H(1-|x'|)$ where H is the Heaviside step function whose value is 1 for positive values of its argument and zero otherwise. The second example (II) corresponds to the Gaussian intensity $\exp(-\frac{1}{2}x'^2)/\sqrt{2\pi}$, while the third example (III) is one in which the surface temperature distribution is such that the

---

[12] See Section 1.2.

incident intensity has a magnitude $1/\pi(1+x'^2)$. It will be referred to as quasi-Gaussian. Its form and the resulting stress distribution are very similar in qualitative terms to the Gaussian case, but the solution is very much simpler in form and far easier to investigate analytically or numerically. Such an example, even if unrealistic at first sight, can be very useful in gaining experience, as it is much easier and quicker to investigate it.

The function χ_2 in each case is

(I) $\quad \chi_2' = \dfrac{z'}{2\pi} \times$

$$\times\left\{(x'-1)\arctan\left(\frac{x'-1}{z'}\right)-(x'+1)\arctan\left(\frac{x'+1}{z'}\right)\right\}+\frac{z'^2}{4\pi}\ln\left\{\frac{z'^2+(1+x')^2}{z'^2+(1-x')^2}\right\}$$

(II) $\quad \chi_2' = \text{Re}\left[\dfrac{iz'}{\sqrt{2\pi}}\int_{z'}^{\infty}\exp\{\tfrac{1}{2}(s-ix')^2\}\,\text{erfc}\{\tfrac{1}{\sqrt{2}}(s-ix')\}ds\right]$

(III) $\quad \chi_2' = -\dfrac{z'}{\pi}\arctan\left(\dfrac{x'}{1+z'}\right).$

Differentiation of (8.10) makes it possible to find the stress components in their dimensionless forms. Numerical investigation shows in all three cases that there can be regions of tensile stress. Figure 8.5 shows the shape of each of the three intensity profiles together with their x'-gradients.

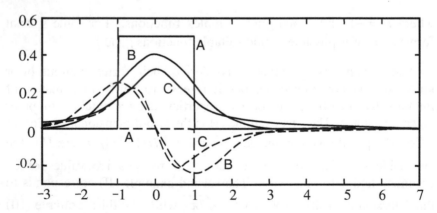

Figure 8.5. Solid lines: A: The top hat profile; B: the Gaussian profile; and C: the quasi-Gaussian profile. The broken lines show the corresponding gradients.

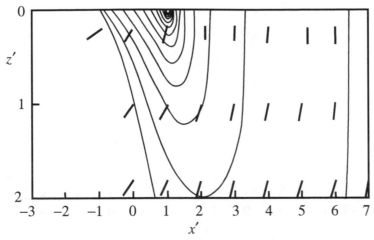

Figure 8.6. Contours of the maximum tensile stress for the top hat case. Contours are at intervals of 0.1 units and the line segments are perpendicular to the direction of the maximum tensile stress.

Figures 8.6 to 8.8 show, respectively, contours of the greatest tensile stress for each of these; the contours are at intervals of 0.1 units. Superimposed are line segments indicating the plane perpendicular to the direction of greatest tensile stress. These directions are perpendicular to the eigenvectors of the stress tensor. The absence of a line segment at the left-hand end of 8.6 and 8.7 indicates that the stress at that point is entirely compressive. These line segments correspond to planes of likely local fracture. The actual direction of fracture is difficult to determine as it will also be guided by the strength of the tractions. When the line of high tension coincides with the possible local plane of fracture, it is probable that the actual direction of fracture will follow this line. If they are roughly in the same direction as the regions of maximum tensile stress, it is plausible to suppose that fracture will be relatively smooth. If their directions are considerably at variance with each other, fracture may well be more jagged.

Comparison of the contour diagrams reveals a number of things. The first feature to notice is that the maximum tensile stress is immediately under the surface thermal layer (and thus at the top of the contour maps), and that it occurs immediately under the region of maximum decrease of intensity.

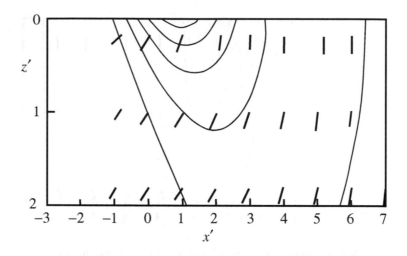

Figure 8.7. Contours of the maximum tensile stress for the Gaussian case. Contours are at intervals of 0.1 units and the line segments are perpendicular to the direction of the maximum tensile stress.

It is clear that

(i) the maximum tensile stress is greater, the greater the gradient of I at this point;

(ii) the scale for variation of stress with depth is the same as the horizontal scale, which is two dimensionless units in these three cases.

In order to achieve a high maximum stress, it is necessary to have a sharp rear edge to the beam. To achieve substantial penetration the horizontal length scale of the beam must be comparable to the depth of material required to be removed. So, for example, to remove material to a depth of 2 cm, a beam of width at least 2 cm in the direction of translation is needed. It is not entirely clear how the beam width should be defined for the purpose. The contours suggest that perhaps it should be the distance between the greatest rate of increase of intensity and the greatest rate of decrease of intensity. Further investigation of the solutions incidentally shows that the former point corresponds to the region of greatest compressive stress.

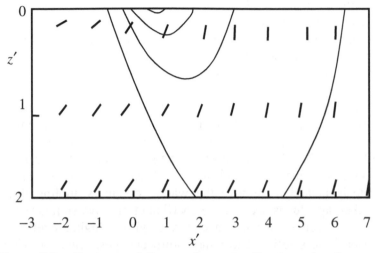

Figure 8.8. Contours of the maximum tensile stress for the quasi-Guassian case. Contours are at intervals of 0.1 units and the line segments are perpendicular to the direction of the maximum tensile stress.

In all three cases the line segments indicating the direction of the line of maximum tension are inclined downward under the irradiated region, rather than under the region that has already completely passed the beam. Perhaps this indicates a tendency for a fracture to break forward from the point at which the tensile stress first reaches its fracture value, removing the surface layer under the beam itself. It is not clear how far this will travel before breaking back to the surface, as the tensile stress decreases quite rapidly in this direction. A more detailed analysis of this aspect of the problem needs a study of fracture mechanics, but it is clear that the general direction tends to be downward.

The diagrams make it clear that the width of the beam determines the depth of penetration, and so it is important to have a broad beam rather than a tightly focused one. The fact that the line segments on the leading edge and bottom of a fracture are not horizontal suggests that the bottom of the region of removed material may have been subject to substantial tensions. These would be caused by cracks that had a tendency to continue downward but were prevented by global considerations. Perhaps this is an explanation for the appearance of micro-cracks in the bottom of the removed region. A possible speculation is that the crack is more likely to return to the surface reasonably well ahead of the incident beam, in a region where the compressive stresses are still relatively low. If that is so, it would also appear preferable to keep the intensity as low as possible. Although the fracturing value of the tensile stresses is reached, it happens relatively near the trailing edge of the incident beam.

In practice, the beam will have a finite width, and it is reasonably clear that the wider this is, the better it will achieve as much removal as possible in a single pass. A question that arises very naturally concerns the role played by the water content of the concrete. A model such as this can give some rough indications. Experiments show that if the concrete is heated to remove the water content, scabbling does not occur. Perhaps this is because the coefficient of expansion is reduced.[13] It has also been shown that a concrete with a higher coefficient of expansion is less resilient to temperature changes. Such a difference results in lower stresses in the interior of the material. It is an idea that can be examined with the help of a model of this kind. Equation (8.11) shows that the stresses in the concrete are directly proportional to the coefficient of expansion, so halving it halves all the stresses. If this brings them below the critical stress, then no damage occurs. Similarly, if either the incident intensity is doubled (with the beam profile unchanged) or the scanning speed is halved, then scabbling should occur, all other things being equal. It is, however, important that the surface should not reach glazing temperature.

The effect of treating the concrete for tests such as these may introduce greater complications, with other parameters of the material being affected. The model still provides an indication of what might occur if the changes could be properly quantified. Clearly, the specific values of the properties of the concrete become critical in determining the optimum values of such things as the incident intensity and scanning speed. A mathematical model such as this one allows a means by which such effects can be studied.

8.3. STRESSES IN A METAL WORKPIECE

The underlying theory for the stresses in a semi-infinite metal work-piece is exactly the same as for the problem of the scabbling of concrete studied in the last section.[14] The only difference is that it is no longer

---

[13] Neville, 1995.

[14] E.g., Postaciolu et al., 1997;Tsuji et al., 1994.

true that the thermal layer is thin compared to the region affected by the induced stresses. The solution for ξ can nevertheless be separated into two parts. One part is directly expressed in terms of the temperature distribution, $\nabla\chi$ in the notation of Appendix 2. The other part satisfies the equations of the theory of linear elasticity without the thermal term, but must satisfy a boundary condition derived from χ.[15]

8.4 CONCLUSION

The example of the scabbling of concrete, studied at length in this final chapter, shows particularly clearly the way in which the act of trying to construct a mathematical model of a process can help to clarify perceptions of what is involved. Thus, it may provide a guide as to what is important and what, perhaps, is not. Finally, it poses questions about the process itself, and in doing so can suggest new directions for investigation which may, in turn, lead to yet more new insights into the problem.

---

[15] In the two-dimensional solution in Section 8.2, the corresponding quantities are χ_1 and χ_2, respectively, for the two parts of the solution.

APPENDIX 1

VALUES OF MATERIAL PROPERTIES

Some typical values of the properties of a number of materials are included. They are intended as no more than a guide, and any reader who wishes for good numerical agreement between a model and experiment should try to obtain the best values possible for the particular material under study. Orders of magnitude, however, especially of dimensionless parameters such as the Péclet number, are frequently invaluable guides to the nature of the approximations that are appropriate in the construction of an analytical model. Sources include Lide and Frederikse (1997), Neville (1995), and Smithells (1962).

Concrete			
Melting temperature	T_M	1573	K
Thermal conductivity (solid)	λ_S	0.8	$\mathrm{W\ m^{-1}\ K^{-1}}$
Thermal diffusivity (solid)	κ_S	10^{-6}	$\mathrm{m^2\ s^{-1}}$
Specific heat (solid)	c_{pS}	500	$\mathrm{J\ kg^{-1}\ K^{-1}}$
Density	ρ	1600	$\mathrm{kg\ m^{-3}}$
Coefficient of linear expansion	α_L α	2×10^{-6}	$\mathrm{K^{-1}}$
Young's modulus	E	3.10×10^{10}	Pa
Poisson's ratio	ν	0.2	–
Lamé constants	λ_L	1.29×10^{10}	Pa
	μ_L	0.86×10^{10}	Pa
Compressive strength		4×10^7	Pa
Tensile strength		4×10^6	Pa

		Aluminium (at 1 bar)	
Surface tension (with air)	γ	0.914	N m^{-1}
Latent heat of vaporization	L_V	10.8×10^6	J kg^{-1}
Boiling temperature	T_B	2740	K
Thermal conductivity (liquid)	λ_L	103	W m^{-1} K^{-1}
Thermal diffusivity (liquid)	κ_L	3.51×10^{-5}	m^2 s^{-1}
Specific heat (liquid)	c_{pL}	1087	J kg^{-1} K^{-1}
Viscosity (molten state)	μ	1.05×10^{-3}	kg m^{-1} s^{-1}
Kinematic viscosity (molten state)	ν	2.85×10^{-7}	m^2 s^{-1}
Prandtl number	Pr	0.0081	–
Latent heat of melting	L_M	3.98×10^5	J kg^{-1}
Melting temperature	T_M	931.7 (Pure) 838 (Alloy 360)	K
Thermal conductivity (solid)	λ_S	237 (Pure) 150 (Alloy 360)	W m^{-1} K^{-1}
Thermal diffusivity (solid)	κ_S	9.75×10^{-5}	m^2 s^{-1}
Specific heat (solid)	c_{pS}	900	J kg^{-1} K^{-1}
Density	ρ	2700 (Pure) 6648 (Alloy 360)	kg m^{-3}
Coefficient of linear expansion	α_L	21.0×10^{-6} (Alloy 360)	K^{-1}
Young's modulus	E	70×10^9 (Alloy 360)	Pa
Tensile strength		325×10^6 (Alloy 360)	Pa

Iron and Stainless Steel - 304 (at 1 bar)			
Surface tension	γ	1.872	N m^{-1}
Latent heat of vaporization	L_V	6.07×10^6	J kg^{-1}
Boiling temperature	T_B	2999	K
Thermal conductivity (liquid)	λ_L	32.7	W m^{-1} K^{-1}
Thermal diffusivity (liquid)	κ_L	0.551×10^{-5}	m^2 s^{-1}
Specific heat (liquid)	c_{pL}	824	J kg^{-1} K^{-1}
Viscosity (molten)	μ	2.25×10^{-3}	kg m^{-1} s^{-1}
Kinematic viscosity (molten)	ν	2.85×10^{-7}	m^2 s^{-1}
Prandtl number	Pr	0.052	–
Melting latent heat	L_M	2.67×10^5	J kg^{-1}
Melting temperature	T_M	1813 (Ingot iron) 1698 (Stainless steel)	K
Thermal conductivity (solid)	λ_S	70 (Ingot iron) 15 (Stainless steel)	W m^{-1} K^{-1}
Thermal diffusivity (solid)	κ_S	2.13×10^{-5}	m^2 s^{-1}
Specific heat (solid)	c_{pS}	450	J kg^{-1} K^{-1}
Density	ρ	7860 (Ingot iron) 7900 (Stainless steel)	kg m^{-3}
Coefficient of linear expansion	α_L	11.7×10^{-6} (Ingot iron) 17.3×10^{-6} (Stainless steel)	K^{-1}
Young's modulus	E	205×10^9 (Ingot iron) 195×10^9 (Stainless steel)	Pa
Tensile strength		55×10^7 (Stainless steel)	Pa

Lead (at 1 bar)			
Latent heat of vaporization	L_V	0.861×10^6	$J\,kg^{-1}$
Boiling temperature	T_B	2013	K
Thermal conductivity (liquid)	λ_L	6.37	$W\,m^{-1}\,K^{-1}$
Thermal diffusivity (liquid)	κ_L	0.407×10^{-5}	$m^2\,s^{-1}$
Specific heat (liquid)	c_{pL}	138	$J\,kg^{-1}\,K^{-1}$
Viscosity (molten state)	μ	1.7×10^{-3}	$kg\,m^{-1}\,s^{-1}$
Kinematic viscosity (molten state)	ν	1.5×10^{-7}	$m^2\,s^{-1}$
Prandtl number	Pr	0.036	–
Latent heat of melting	L_M	0.23×10^5	$J\,kg^{-1}$
Melting temperature	T_M	600	K
Thermal conductivity (solid)	λ_S	34.6	$W\,m^{-1}\,K^{-1}$
Thermal diffusivity (solid)	κ_S	2.35×10^{-5}	$m^2\,s^{-1}$
Specific heat (solid)	c_{pS}	130	$J\,kg^{-1}\,K^{-1}$
Density	ρ	11340	$kg\,m^{-3}$
Young's modulus	E	1.5×10^{10}	Pa

Titanium (at 1 bar)			
Surface tension (with air)	γ	1.650	N m^{-1}
Boiling temperature	T_B	3560	K
Latent heat of melting	L_M	4.187×10^5	J kg^{-1}
Melting temperature	T_M	1941 (Pure) 1943 (Commercial)	K
Thermal conductivity (solid)	λ_S	20 (Pure) 180 (Commercial)	W m^{-1} K^{-1}
Thermal diffusivity (solid)	κ_S	0.86×10^{-5}	m^2 s^{-1}
Specific heat (solid)	c_{pS}	523	J kg^{-1} K^{-1}
Density	ρ	4510 (Pure) 4500 (Commecial)	kg m^{-3}
Coefficient of linear expansion	α_L	8.5×10^{-6} (Commercial) 8.6×10^{-6} (Pure)	K^{-1}
Young's modulus	E	110×10^9 (Commercial)	Pa
Tensile strength		$330 - 500 \times 10^6$ (Commercial)	Pa

Water and ice (at 1 bar)			
Surface tension (with air)	γ	0.073	N m^{-1}
Latent heat of vaporization	L_V	2.25×10^6	J kg^{-1}
Boiling temperature	T_B	373	K
Thermal conductivity (water)	λ_L	0.6	W m^{-1} K^{-1}
Thermal diffusivity (water)	κ_L	1.43×10^{-7}	m^2 s^{-1}
Specific heat (water)	c_{pL}	4200	J kg^{-1} K^{-1}
Viscosity (water)	μ	1.0×10^{-3}	kg m^{-1} s^{-1}
Kinematic viscosity (water)	ν	1.0×10^{-5}	m^2 s^{-1}
Prandtl number	Pr	70	–
Density	ρ_L	1000	kg m^{-3}
Latent heat of melting	L_M	0.33×10^6	J kg^{-1}
Melting temperature	T_M	273	K
Thermal conductivity (ice)	λ_S	2.2	W m^{-1} K^{-1}
Thermal diffusivity (ice)	κ_S	1.13×10^{-6}	m^2 s^{-1}
Specific heat (ice)	c	2120	J kg^{-1} K^{-1}
Density	ρ_S	920	kg m^{-3}

APPENDIX 2

ELASTIC GREEN'S FUNCTIONS FOR A SEMI-INFINITE DOMAIN

In this appendix, the symbols λ and μ refer throughout to the Lamé constants, written λ_L and μ_L elsewhere in the text.

In order to solve Equation (8.8) with the boundary conditions (8.9), it is helpful to write $\xi = \zeta + \nabla\chi$ where

$$\nabla^2\chi = \frac{3\lambda + 2\mu}{\lambda + 2\mu}\alpha(T - T_0) \text{ or } \chi = \frac{3\lambda + 2\mu}{\lambda + 2\mu}\frac{\alpha\kappa}{U}\int_{x'=-\infty}^{x}(T - T_0)dx$$

on the assumption that $\lim_{x \to -\infty} T = T_0$. In that case, and because of the very large value of Young's modulus for concrete, which shows that the inertial terms in Equation (2.72) may be neglected, ζ satisfies the equation

$$(\lambda + \mu)\nabla\nabla.\zeta + \mu\nabla^2\zeta = 0 \tag{A1}$$

with the boundary conditions

$$\lambda\nabla.\zeta\delta_{i3} + \mu\left(\frac{\partial\zeta_i}{\partial x_3} + \frac{\partial\zeta_3}{\partial x_i}\right) = -\begin{pmatrix} 2\mu_L\chi_{,13} \\ 2\mu_L\chi_{,23} \\ \lambda\nabla^2\chi + 2\mu\chi_{,33} - (3\lambda + 2\mu)\alpha(T - T_0) \end{pmatrix}$$

$$= -\frac{2\mu(3\lambda + 2\mu)\alpha\kappa}{(\lambda + 2\mu)U}\begin{pmatrix} \dfrac{\partial T}{\partial z} \\ \displaystyle\int_{x'=-\infty}^{x}\dfrac{\partial^2 T}{\partial y\partial z}dx' \\ \dfrac{\partial T}{\partial x} + \displaystyle\int_{x'=-\infty}^{x}\dfrac{\partial^2 T}{\partial y^2}dx' \end{pmatrix} \equiv \tau(x, y) \tag{A2}$$

269

at $z = 0$. The use of subscripts preceded by a comma indicates differentiation with respect to the corresponding component of the co-ordinate vector.

Because of the thin layer approximation that is appropriate in this problem, the right-hand side of the z-component of the boundary condition can be taken as zero. Consequently, from (A2) the approximate boundary condition at $z = 0$ is

$$\lambda \nabla \cdot \zeta \delta_{i3} + \mu \left(\frac{\partial \zeta_i}{\partial x_3} + \frac{\partial \zeta_3}{\partial x_i} \right) = \frac{2\mu(3\lambda + 2\mu)\alpha\kappa}{(\lambda + 2\mu)U\lambda_c} \left(\begin{array}{c} I \\ \int_{x'=-\infty}^{x} \frac{\partial I}{\partial y} dx' \\ 0 \end{array} \right). \quad \text{(A3)}$$

Use has here been made of the surface boundary condition on the temperature,

$$-\lambda_c \frac{\partial T}{\partial z} \bigg|_{z=0} = I(x, y) \text{, the incident intensity distribution.}$$

Unlike the general thermoelastic problem, it will be seen that it is not necessary to find the temperature distribution in order to solve Equation (A3).

The solution of the equations of the linear theory of elasticity can be expressed in terms of point solutions when the normal stress is specified at the surface $z = 0$ of a semi-infinite domain $z \geq 0$ if it is assumed that all stresses tend to zero at infinity; it first obtained by Boussinesq.[1] The appropriate point source solution is required to have a delta-function behavior for the normal stress at the surface (at a point that is most conveniently chosen to be the origin), and to tend to zero far from the boundary. It can be obtained without undue difficulty by the following method. Construct the Fourier transform in x and y of the problem to be solved. This results in a set of coupled ordinary differential equations. They can, in principle, be solved by hand, but are more conveniently solved using a computer algebra package such as Maple[®2]. When the

---

[1] Landau and Lifshitz, 1959a, p.26.
[2] Registered trademark of Waterloo Maple, Inc., Waterloo, Canada.

boundary conditions at the surface and at infinity have been imposed, the terms can be regrouped into relatively simple forms. The inverse Fourier transforms of some of the terms are readily identifiable, and the remainder can be inferred from these. This procedure lacks rigor, but having obtained the solution it is easily verified (using a computer algebra package for preference) that they do indeed satisfy the equation. The delta function behavior can be checked by direct analytical means backed by the use of a package to obtain the stress components from the displacements. An alternative approach is that described by Landau and Lifshitz.

The stress distribution is then given in $z \geq 0$ by

$$p_{ij}(x,y,z) =$$

$$-\sum_{k=1}^{3} \int_{y'=-\infty}^{y'=\infty} \int_{x'=-\infty}^{x'=\infty} \tau_k(x-x',y-y') P_{ij}^{(k)}(x',y',z) dx' dy'$$

(A4)

and satisfies the boundary condition $-p_{i3}(x,y,0) = \tau_i(x,y)$, for $i = 1..3$; τ is given and is the stress exerted on the medium by its surroundings. Note that the unit normal out of the medium at its surface is $(0,0,-1)$.

If $r = \sqrt{x^2 + y^2 + z^2}$ and $(x,y,z) \equiv (x_1, x_2, x_3)$, the components of the tensor $P_{ij}^{(k)}$ can be obtained by either of the techniques described and are given by

$$P_{i3}^{(k)} = P_{3i}^{(k)} = \frac{3z}{2\pi r^5} x_i x_k \quad (i,k = 1..3),$$

$$P_{11}^{(1)} = \frac{3x^3}{2\pi r^5} + x(y^2 K - L), \quad P_{22}^{(1)} = \frac{3xy^2}{2\pi r^5} + x(x^2 K - 3L),$$

$$P_{12}^{(1)} = P_{21}^{(1)} = \frac{3x^2 y}{2\pi r^5} - y(x^2 K - L),$$

$$P_{11}^{(2)} = \frac{3x^2 y}{2\pi r^5} + y(y^2 K - 3L), \quad P_{22}^{(2)} = \frac{3y^3}{2\pi r^5} + y(x^2 K - L),$$

$$P_{12}^{(2)} = P_{21}^{(2)} = \frac{3y^2x}{2\pi r^5} - x(y^2K - L),$$

$$P_{11}^{(3)} = \frac{x^2}{2\pi r^3}M + N, \; P_{22}^{(3)} = \frac{y^2}{2\pi r^3}M + N, \; P_{12}^{(3)} = P_{21}^{(3)} = \frac{xy}{2\pi r^3}M,$$

$$(A5)$$

where

$$K = \frac{\mu(z+3r)}{2\pi(\lambda+\mu)r^3(z+r)^3}, \quad L = \frac{\mu}{2\pi(\lambda+\mu)r(z+r)^2},$$

$$M = \frac{3z}{r^2} - \frac{\mu(z+2r)}{(z+r)^2(\lambda+\mu)} \quad \text{and} \quad N = \frac{\mu}{2\pi r(\lambda+\mu)(z+r)} - \frac{\mu z}{2\pi r^3(\lambda+\mu)}.$$

$$(A6)$$

They are derived from

$$X^{(1)} = -\frac{1}{4\pi}\left\{\frac{1}{(\lambda+\mu)}\left(\frac{1}{r+z} - \frac{x^2}{r(r+z)^2}\right) + \frac{1}{\mu}\left(\frac{x^2}{r^3} + \frac{1}{r}\right)\right\},$$

$$Y^{(2)} = -\frac{1}{4\pi}\left\{\frac{1}{(\lambda+\mu)}\left(\frac{1}{r+z} - \frac{y^2}{r(r+z)^2}\right) + \frac{1}{\mu}\left(\frac{y^2}{r^3} + \frac{1}{r}\right)\right\},$$

$$Y^{(1)} = X^{(2)} = \frac{xy}{4\pi}\left\{\frac{1}{(\lambda+\mu)r(r+z)^2} - \frac{1}{\mu r^3}\right\},$$

$$\frac{Z^{(1)}}{x} = \frac{Z^{(2)}}{y} = -\frac{1}{4\pi}\left\{\frac{1}{(\lambda+\mu)r(r+z)} + \frac{z}{\mu r^3}\right\},$$

$$\frac{X^{(3)}}{x} = \frac{Y^{(3)}}{y} = \frac{1}{4\pi r}\left\{\frac{1}{(\lambda+\mu)(r+z)} - \frac{z}{\mu r^2}\right\},$$

$$Z^{(3)} = \frac{1}{4\pi r}\left\{-\frac{\lambda+2\mu}{\mu(\lambda+\mu)} - \frac{z^2}{\mu r^2}\right\}.$$

$$(A7)$$

These expressions for $P_{ij}^{(k)}$ are undefined at $z = 0$, but formally in the limit as $z \downarrow 0$,

$$P_{13}^{(1)}(x, y, 0) = P_{31}^{(1)}(x, y, 0) = P_{23}^{(2)}(x, y, 0) = P_{32}^{(2)}(x, y, 0) = P_{33}^{(3)}(x, y, 0) = \delta(x, y),$$

$$P_{11}^{(3)}(x, y, 0) = P_{22}^{(3)}(x, y, 0) = \frac{2\lambda + \mu}{2(\lambda + \mu)} \delta(x, y)$$

$$(A8)$$

while all the rest are zero. Here, $\delta(x, y)$ is the Dirac delta function, which is zero everywhere except at $x = y = 0$ and has unit integral over any two-dimensional domain containing the origin.

BIBLIOGRAPHY

Abramowitz, M. and Stegun, I. *Handbook of Mathematical Functions.* Dover, New York, 1965.

Akhter, R., Davis, M. P., Dowden, J. M., Ley, M., Kapadia, P., and Steen, W. M. A method for calculating the fused zone profile of laser keyhole welds. *J.Phys D: Appl Phys,.* 22, 23-29, 1989.

Andrews, J. G. Mathematical models in welding. *Bull. Inst. Maths. Applics.,* 15.10, 250-53, 1979.

Andrews, J. G. and Atthey, D. R. On the motion of an intensely heated evaporating boundary. *J. Inst. Maths. Applics.,* 15, 59-72, 1975.

Andrews, J. G. and Atthey, D. R. Hydrodynamic limit to penetration of a material by a high-power beam. *J. Phys. D: Appl. Phys.,* 9, 2181-2194, 1976.

Arata, Y. Challenge of laser advanced materials processing. *Proc. Conf. Laser Advanced Material Processing LAMP'87.* Osaka: High Temperature Society of Japan, 1987, 3-11.

Arata, Y. and Miyamoto, I. *Technocrat* 11, pp.33-42, 1978.

Arata, Y., Abe, N., and Oda, T. Beam hole behaviour during laser beam welding. *Proc. ICALEO'83.* Laser Institute of America, Orlando, 1983, 38, 59-66.

Arata, Y., Maruo, H., Miyamoto, I., and Takeuchi, S. Dynamic behaviour of laser welding and cutting. *Proc. 7th. Int. Conf. Electron and Ion Beam,Science and Technology.* pp.111-128, 1976.

Ashby, M. F. and Easterling, K .E. The transformation hardening of steel surfaces by laser beams. *Acta Metall.* 32, 1935-48, 1984.

Atthey, D. R. A finite difference scheme for melting problems. *J. Inst. Math. Applics.,* 13.3, 353-366, 1974.

Avilov, V. V., Vicanek, M., and Simon, G. Thermal diffusion in laser beam welding of metal. *J. Phys. D: Appl. Phys.* 29, 1146-1156, 1996.

Basalaeva, M. A. and Bashenko, V. V. The movement of metal in the weld pool in electron beam welding. *Weld. Prod.,* 24.3, 1-3, 1977.

Batchelor, G. K. *An Introduction to Fluid Dynamics.* Cambridge University Press, Cambridge, 1967.

Berger, J. M. Absorption coefficients for free-free transitions in a hydrogen plasma. *Astrophys. J.,* 124, 550-554, 1956.

Berger, P. In *Proceedings of the Twelfth Meeting on Mathematical Modeling of Materials Processing with Lasers,* Kaplan A. and Schuöcker, D., Eds., ELA/ARGELAS, Vienna, 1997.

Beyer, E., Maischner, D., and Kratzsch, Ch. A neural network to analyze plasma fluctuations with the aim to determine the degree of full penetration in laser welding. *Proc. ICALEO '94.* Laser Institute of America, Orlando, 1995, 51-57.

Boley, B. B. and Weiner, J. H. *Theory of Thermal Stresses.* Dover, New York, 1960.

Carslaw, H. S. and Jaeger, J. C. *Conduction of Heat in Solids.* Clarendon Press, Oxford, Second edition, 1959.

Chandrasekhar, S. Stochastic problems in physics and astronomy. *Rev. Mod. Phys.* 15.1, 2-87, 1943.

Chandrasekhar, S. and Münch, G. The theory of fluctuations in brightness of the Milky Way, I-V. *Astrophys. J.,* 112, 380-392, 1950; *Astrophys. J.,* 112, 393-398, 1950; *Astrophys. J.,* 114, 110-122, 1951; *Astrophys. J.,* 115, 94-102, 1952; *Astrophys. J.,* 115, 103-123, 1952.

Chapman, A. England's Leonardo: Robert Hooke (1635-1703) and the art of experiment in Restoration England. *Proc. Royal Institution of Great Britain,* 1996, 67, 239-275.

Cline, H. E. and Anthony, T. R. Heat treating and melting material with a scanning laser or electron beam. *J. Appl. Phys.* 48, 3895-3900, 1977.

Clucas, D. A. V., Steen, W. M., Ducharme, R., Kapadia, P. D., and Dowden, J.M. A mathematical model of the flow within the keyhole during laser welding. *Proc. ICALEO'95,* Laser Institute of America, Orlando, USA, 1996, 80, 989-998.

Colla, T. J., Vicanek, M., and Simon, G. Heat transport in melt flowing past the keyhole in deep penetration welding. *J. Phys. D: Appl. Phys.* 27, 2035-40, 1994.

Collins, R., Pemberton, J., Pemberton, S. J. D., and Matthews, J. A. Output velocity distribution of a Langevin system with a random binary input. *Eur. J. Phys.* 9, 312-318, 1988.

Cram, L. E. Statistical evaluation of radiative power losses from thermal plasmas due to spectral lines. *J Phys D: Appl Phys.* 18, 401-411, 1985.

Davis, M. *Fluid Dynamical Models of Laser Welding.* Thesis for the degree of PhD, University of Essex, Colchester, 1983.

de Groot, S. R. *Thermodynamics of Irreversible Processes.* North Holland, Amsterdam, 1952.

Dilawari, A. H., Eagar, T. W., and Szekely, J. An analysis of heat flow phenomena in electroslag welding. *Welding J.,* 57.1, 24s-30s, 1978.

Dowden, J. M. and Kapadia, P. D. A mathematical model of the chevron-like wave pattern on a weld piece. *Proc. ICALEO'96,* Laser Institute of America, LIA, Orlando, 1997, 81, B96-B105.

Dowden, J. M. and Kapadia, P. D. Acoustic oscillations in the keyhole in laser welding. *Int. J. Joining of Materials,* 10, No. 1/2, 25-32, 1998.

Dowden, J. M. and Kapadia, P. D. Mathematical investigation of the penetration depth in keyhole welding with continuous CO_2 lasers. *J. Phys. D: Appl. Phys., 28,* 2252-2261, 1995.

Dowden, J. M. and Kapadia, P. D. The instabilities of the keyhole and the formation of pores in the weld in laser keyhole welding. *Proc. ICALEO'95,* Laser Institute of America, Orlando, USA, 1996, 80, 961-968.

Dowden, J. M., Davis, M., and Kapadia, P. D. Some aspects of the fluid dynamics of laser welding. *J. Fluid Mech.,* 126, 123-146, 1983.

Dowden, J. M., Ducharme, R., and Kapadia, P. D. Time-dependent line and point sources: a simple model for time-dependent welding processes. *Lasers in Engineering,* 7 (3-4), 215-228, 1998.

Dowden, J. M., Ducharme, R., Kapadia, P. D., and Clucas, A. A mathematical model for the penetration depth in welding with continuous CO_2 lasers. *Proc. ICALEO'94.* Laser Institute of America, Orlando, 1995a, 79, 451-460.

Dowden, J. M., Kapadia, P. D., and Davis, M. The fluid dynamics of laser welding. In Cheremisinoff, N. P., Ed., *Encyclopedia of Fluid Mechanics,* Vol .VI, Gulf, Houston, 1987.

Dowden, J. M., Kapadia, P. D., and Ducharme, R. An analytical model of deep penetration welding of metals with a continuous CO_2 laser. *Int. J. Joining of Materials,* 7, 54-62, 1995b.

Dowden, J. M., Kapadia, P. D., and Ducharme, R. Temperature in the plume in penetration welding with a laser. *Transport Phenomena in Materials Processing and Manufacturing ASME 1994,* American Society of Mechanical Engineers, New Yark, 1994, HTD-Vol. 280, 101-106.

Dowden, J. M., Kapadia, P. D., and Fenn, R. Space charge in plasma arc welding and cutting. *J. Phys. D: Appl. Phys.* 26, 1215-1223, 1993.

Dowden, J. M., Kapadia, P. D., and Postacioğlu, N. An analysis of the laser-plasma interaction in laser keyhole welding. *J.Phys. D: Appl. Phys.* 22, 741-749, 1989.

Dowden, J. M., Kapadia, P. D., and Sibold, D. Mathematical modelling of laser welding *Int. J. Joining Mater*, 3, 73-8, 1991.

Dowden, J. M., Kapadia, P. D., Clucas, A., Ducharme, R., and Steen, W. M. Laser welding: On the relation between fluid dynamic pressure and the formation of pores in laser keyhole welding. *J. Laser Applications*, 8, 183-190, 1996.

Dowden, J. M., Wu, S.C., Kapadia, P. D., and Strange, C. M. Dynamics of the vapour flow in the keyhole in penetration welding with a laser. *J. Phys. D: Appl. Phys.*, 24, 519-532, 1991.

Ducharme, R., Kapadia, P. D., and Dowden, J. M. A mathematical model of the defocusing of laser light above a work piece in laser material processing. *Proc. ICALEO'92*, Laser Institute of America, Orlando, 1993, 75, 187-197.

Ducharme, R., Kapadia, P. D., and Dowden, J. M. The collapse of the keyhole in the laser welding of materials. *Proc. ICALEO'93*, Laser Institute of America, Orlando, 1994a, 77, 177-183.

Ducharme, R., Kapadia, P. D., Dowden, J. M., Hilton, P., Riches S., and Jones, I. A. An analysis of the laser material interaction in the welding of steel using a CW CO laser. *Proc. ICALEO'96*, Laser Institute of America, Orlando, 1997, 81, D10-D20.

Ducharme, R., Kapadia, P. D., Dowden, J. M., Williams, K., and Steen, W. M. An integrated mathematical model for the welding of thick sheets of metal with a continuous CO_2 laser. *Proc. ICALEO'93*, Laser Institute of America, Orlando, 1994b, 77, 97-105.

Ducharme, R., Kapadia, P. D., Lampa, C., Ivarson, A., Powell J., and Magnusson, C. A point and line source analysis of the laser material interaction in hyperbaric keyhole laser welding. *Proc. ICALEO'95*, Laser Institute of America, Orlando, USA, 1996, 80, 1018-1027.

Ducharme, R., Williams, K., Kapadia, P. D., Dowden, J. M., Steen, W. M., and Glowack, M. The laser welding of thin metal sheets: an integrated keyhole and weld pool model with supporting experiments. *J. Phys. D: Appl. Phys.*, 27, 1619-27, 1994c.

Duley, W. W. CO_2 *Lasers: Effects and Applications*. Academic Press, London, 1976.

Eliezer, S., Ghatak, A., and Hora, H. *Theory and Applications: Equations of State.* Cambridge University Press, Cambridge, 1986.

Emsley, J. *The Elements.* Clarendon Press, Oxford, 1998.

Farson, D. F., Fang, K. S., and Kern, J. Intelligent laser welding control. *Proc. ICALEO '91.* Laser Institute of America, Orlando, 1992, 104-112.

Ferlito, C. and Riches, S .T. The influence of plasma control gases on laser weld quality in C-Mn steel. Cooperative Research Programme, Report 477, 1993. The Welding Institute, Cambridge, 1993.

Finke, B. R. and Simon, G. On the gas kinetics of laser-induced evaporation of metals. *J. Phys. D: Appl. Phys.*, 23, (1) 67-74, 1990.

Finke, B. R., Kapadia, P., and Dowden, J. M. Fundamental plasma gas model for energy transfer in laser material processing. *J. Phys. D: Appl. Phys.*, 23, 643-654, 1990.

Frenkel, J. *Kinetic Theory of Liquids.* Dover Publications, New York, 1972.

Gellert, M., Kapadia, P. D., Ducharme, R., Dowden, J. M., and Simon, G. Investigation of the migration and growth of bubbles in liquid metals in high temperature gradients. Presented at the meeting of the EU194 Mathematical Modelling Group, Jan. 1995, in Innsbruck, Austria, 1995.

Gillies, B. *The Double Free Boundary Value Problem of Laser Welding of Thin Sheets at Medium Speeds.* Thesis for the degree of Ph.D., Heriot-Watt University, Edinburgh, 2000.

Goldak, J. A., Burbidge, G., and Bibby, M. J. Predicting micro structure from heat flow calculations in electron beam welded eutectoid steels. *Can. Met. Quart.*, 9.3, 459-66, 1970.

Gouveia, H. *The Coupling Mechanism in the CO_2 Laser Welding of Copper.* Thesis for the degree of Ph.D., Cranfield University, 1994.

Gouveia, H., Richardson, I., Kapadia, P., Dowden, J. M., and Ducharme, R. The laser welding of copper using the integrated keyhole and weld pool model and continuous CO_2 and Nd:YAG lasers. *Proc. ICALEO'94*, Laser Institute America, Orlando, 1995, 79, 480-489.

Gratzke, U. and Simon, G. Laser-induced oxidation process of tungsten. *J. Phys. D: Appl. Phys.*, 24, 827-834, 1991.

Gratzke, U., Kapadia, P. D., and Dowden, J. M. Theoretical approach to the humping phenomenon in welding processes. *J. Phys. D: Appl. Phys.* 25, 1640-47, 1992.

Heiple, C. R. and Roper, J. R. *Welding J.* 61, 97s-102s, 1982.

Henry, P., Chande, T., Lipscombe, K., Mazumder, J., and Steen, W. M. *Proc. ICALEO'82*, Laser Institute of America, Orlando, USA, 1982, paper 4B-2.

Homann, F. Der Einfluß großer Zähligkeit bei der Strömung um der Zylinder und die Kugel. *ZAMM*, 16, 153, 1936.

Hooke, R. *A Description of Helioscopes and some other Instruments*, London, 1676, p.32, item 9.

Hooke, R. *De Potentia Restitutiva*, London, 1678, p.5.

Hopkins, J. A., McCay, T. D., McCay, M. H., and Eraslan, A. Transient predictions of CO_2 spot welds in Iconel 718. *Proc. ICALEO'93*, Laser Institute of America, Orlando, 1994, 77, 106-11.

Hügel, H. *Strahlwerkzeug Laser*. Teubner, Stuttgart, 1994.

Hughes, T. P. *Plasmas and Laser Light*. Adam Hilger, London, 1975.

Hunter, S. C. *Mechanics of Continuous Media*. Ellis Horwood, New York, 1983.

Ishimaru, A. *Wave Propagation and Scattering in Random Media*, Vol. I. John Wiley Sons, New York, 1978.

Ishizaki, K. Dynamic Surface Tension and Surface Energy Theory on the Heat Transfer and Penetration in Arc Welding. *IIW Doc*. 212-719-89, 1989.

Jeffreys, H. *Cartesian Tensors*. Cambridge University Press, Cambridge, 1957.

Johnston, E. P., Shannon, G., Steen, W. M., Jones, D. R., and Spencer, J. T. Evaluation of high-powered lasers for a commercial laser concrete scabbling (large-scale ablation) system. *Proc. ICALEO'98*, Laser Institute of America, Orlando, 1999, 85, A210-218.

Kapadia, P. D. and Dowden, J. M. Some electrical effects in the keyhole plasma in deep penetration CW CO_2 laser welding, *Applied Surface Science*, 103 No. 4, 240-242, 1996.

Kapadia, P. D., Solana, P., and Dowden, J. M. Stochastic model of the deep penetration laser welding of metals. *J. Laser Applics.*, 10 No.4 170-73, 1998.

Kapadia, P., Ducharme, R., and Dowden, J. M. A mathematical model of ablation in the keyhole and droplet formation in the plume in deep penetration laser welding. *Proc. ICALEO'96*, Laser Institute of America, Orlando, 1997, 81, B106-B115.

Kaplan A. *Modellrechnung und numerische Simulation von Absorption, Wärmeleitung und Strömung des Laser-Tiefschweissens*. Ph.D. Thesis. Technische Universität, Vienna, 1994.

Kaplan, A. Heat transfer during laser material processing. In *Proc. of the Twelfth Meeting on Mathematical Modeling of Materials Processing with Lasers (M4PL 12, Igls, Austria)*, Department of Laser Technology, Vienna University of Technology, Vienna, March 1997, Paper 5 of Section 2.

Katayame, S., Seto, N., Mizutani, M., and Matsunawa, A. Formation mechanism of porosity in high power YAG laser welding. *Proc. ICALEO 2000*, Laser Institute of America, Orlando, 2000, 89, C16-25.

Kaye, S. A., Delph, A .G., Hanley, E., and Nicholson, C. J. Improved welding penetration of a 10 kW industrial laser. Fourth Int. Symp. on Gas Flow & Chem. Lasers. Stresa, 1983.

Keene, B. J. The Effects of Thermocapillary Flow on Weld Pool Profile. *NPL Report* DMA(A) 167, 1988.

Kern, M., Berger, P., and Hügel, H. Magneto-fluid dynamic control of seam quality in CO_2 laser beam welding. *Welding J.*, 79.3, 72s-78s, 2000.

Klein, T., Paulini, J., and Simon, G. Time-resolved description of cathode spot development in vacuum arcs. *J. Phys. D: Appl. Phys.*, 27, 1914-21, 1994.

Klein, T., Vicanek, M., and Simon, G. Forced oscillations of the keyhole in penetration laser beam welding. *J. Phys. D: Appl. Phys.* 29, 322-32, 1996.

Klein, T., Vicanek, M., Kroos, J., Decker, I., and Simon, G. Oscillations of the keyhole in penetration laser beam welding. *J. Phys. D: Appl. Phys*. 27, 2023-30, 1994.

Klemens, P. G. Heat balance and flow conditions for electron beam welding and laser welding. *J. Appl. Phys.*, 47, 2165-74, 1976.

Kotecki, D. J., Cheever, D. L., and Howden, D. G. Mechanism of ripple formation during weld solidification. *Welding J.*, 58.8, 386s-91s, 1972.

Kreyszig, E. *Advanced Engineering Mathematics.* 7th Edition. Wiley, New York, 1993.

Kroos, J., Gratzke, U., and Simon, G. Towards a self-consistent model of the keyhole in penetration laser-beam welding. *J. Phys. D: Appl. Phys.*, 26, 474-480, 1993a.

Kroos, J., Gratzke, U., and Simon, G. Towards a self-consistent model of the keyhole in laser beam welding. *J. Phys. D: Appl. Phys*. 26, 474-80, 1993b.

Kroos, J., Gratzke, U., Vicanek M., and Simon, G. Dynamic behaviour of the keyhole in laser welding. *J. Phys. D: Appl. Phys.* 26, 481-6, 1993.

La Rocca, A. V. Laser applications in manufacturing. *Sci. Am.*, 264, 80-87, 1982.

Lamb, H. *Hydrodynamics* (sixth edition). Cambridge University Press, Cambridge, 1932.

Landau, L. D. and Lifshitz, E. M. *Theory of Elasticity.* Pergamon, Oxford, 1959a.

Landau, L. D. and Lifshitz, E. M. *Fluid Mechanics.* Pergamon, Oxford, 1959b.

Landau, L. D. and Lifshitz, E. M. *Statistical Physics,* Part 1, 3rd Edition. Pergamon, Oxford, 1980.

Li, L., Brookfield, D. J., and Steen, W. M. Plasma charge sensor for in-process, non-contact monitoring of the laser welding process. *Measurement Science and Technology*, 7.4, 615-626, 1996.

Li, L., Modern, P. J., and Steen, W. M. Concrete decontamination by laser surface treatment. *European Patent* No. 94307937.6.

Li, L., Modern, P. J., and Steen, W. M. Laser surface modification techniques for potential decommissioning applications. *Proc. RECOD '94*, 4th. Conference on Nuclear Fuel Reprocessing and Waste management, 24th.- 28th. April 1994, London, Vol. III, 427-440, 1994.

Liboff, R. L. *Kinetic Theory: Classical, Quantum and Relativistic Descriptions.* Prentice Hall International Inc., Englewood Cliffs NJ, USA., 1990.

Lide, D. R. and Frederikse, H. P. R., Eds., *Handbook of Physics and Chemistry*, 78th Edition. CRC Press, Boca Raton, USA, 1997.

Magee, J., Okon, P., and Dowden, J. M. The relation between spot size and penetration depth in laser welding. IIW Doc. IV-764-2000/212-975-00. International Institute of Welding, Paris, 2000.

Malmuth, N. D. Temperature field of a moving point-source with change of state. *Int. J. Heat Mass Transfer*, 19, 349-54, 1976.

Maple V. MathSoft Inc, Cambridge, Mass., USA. Maple is a registered trademark of Waterloo Maple Inc., Waterloo, Canada.

Mara, G. L. Penetration mechanisms of electron beam welding and the spiking phenomenon. *Welding J.*, 53, 246s-51s, 1974.

Martin, D. H. and Bowen, J. W. Long-wave optics. *IEEE Trans Microwave Theory and Techniques*, 41.10, 1676-90, 1993.

Matsunawa, A. *Possible Motive Forces of Liquid Motion in Laser Weld Pool.* IIW Doc. IV-770-2000/212-917-00. International Institute of Welding, Paris, 2000.

Matsunawa, A. and Ohnawa, T. Beam-plume interaction in laser materials processing. *Trans. Jap. Welding Res. Inst.* 2000, 20.1, 9-15.

Matsunawa, A. and Semak, V. The simulation of front keyhole wall dynamics during laser welding. *J. Phys. D: Appl. Phys.*, 30, 798-809, 1997.

Mazumder, J. In *Laser Materials Processing*, Bass, M., Ed., North Holland, Amsterdam, 1983.

Mazumder, J. and Steen, W. M. Heat transfer for cw laser material processing. *J. Appl. Phys.* 51, 941-7, 1980.

McLachlan, N. W. *Theory and Application of Mathieu Functions,* Clarendon Press, Oxford, 1947.

Megaw, J. H. P. C. and Kaye, A. S. *Surface modification and welding by laser.* UKAEA Culham Laboratory Report CLM-R185, 1978.

Melcher, J. R. *Continuum Electromechanics.* The MIT Press, Cambridge, MA, 1981.

Melcher, J. R., Sachar, K. S., and Warren, E. P. Overview of electrostatic devices for control of submicrometer particles. *Proc. IEEE* 65, 1977, 1659.

Metiu, H., Kitahara, K., and Ross, J. *J. Chem. Phys.*, 64, 292, 1976.

Metiu, H., Kituhara, K., and Ross, J. Statistical mechanics: the theory of the kinetics of phase transitions in: *Fluctuation Phenomena.* Montroll, E. N. and Lebowitz, J. L., Eds., North Holland Pub. Co., 259-321, 1987.

Miyazaki, T. and Giedt, W. H. Heat transfer from an elliptical cylinder through an infinite plate applied to electron beam welding. *J. Heat Mass Transfer*, 25.6, 807-814, 1982.

Montroll, E. W. *Statistical Mechanics.* S. K. Rice, Freed, K. F., and J.C. Light, Eds., Univ. Chicago Press, Chicago, 1972.

Moore, C. E. *Atomic Energy Levels.* National Bureau of Standards Circular 467, Washington, 1949.

Morse, P. M. and Feshbach, H. *Methods of Theoretical Physics.* McGraw-Hill, New York, 1953.

Muskhelishvili, N. I. *Some Basic Problems of the Mathematical Theory of Elasticity* (translated from the third Russian edition by T. R. M. Radok), Noordhof, Groningen, The Netherlands, 1953.

Neville, A. M. *Properties of Concrete* (4th. edn.). Addison Wesley Longman, Harlow, 1995.

Noller, F. The stationary shapes of vapor cavity and melt zone in eb-welding *3rd Int. Conf. on Welding and Melt, Electrons and laser Beam* 89-97, 1983.

Nowacki, W. *Thermoelasticity*. Pergamon, Oxford, 1986.

Ol'shanskii, N. A., *et al.* Movement of molten metal during electron beam welding. *Weld. Prod.*, 12, 20-23, 1974.

Paulini, J. and Simon, G. A theoretical lower limit for laser power in laser-enhanced arc-welding. *J. Phys D: Appl. Phys.*, 26, 1523-27, 1993.

Paulini, J., Klein, T., and Simon, G. Thermo-field emission and the Nottingham effect. *J. Phys D: Appl. Phys.*, 26, 1310-15, 1993.

Paulini, J., Simon, G., and Decker, I. Beam deflection in electron-beam welding by thermoelectric eddy currents. *J. Phys D: Appl. Phys.*, 23, 486-95, 1990.

Peel, D. A. Some moving boundary problems in the steel industry. In *Moving Boundary Problems in Heat Flow and Diffusion*, Ockendon, J. R. and Hodgkin, W. R., Eds., Clarendon Press, Oxford, 5-18, 1975.

Pirri, A. N., Root, R. G., and Wu, P. K. S. Plasma energy transfer to metal surfaces irradiated by pulsed lasers. *AAIAJ*, 16, 1296-1304, 1978.

Postacioğlu, N., Kapadia, P. D., and Dowden, J. M. A theoretical model of thermo-capillary flows in laser welding. *J. Phys. D: Appl. Phys.*, 24: pp.15-20, 1991a.

Postacioğlu, N., Kapadia, P. D., and Dowden, J. M. Capillary waves on the weld pool in penetration welding with a laser. *J. Phys. D: Appl. Phys.*, 22, 1050-1061, 1989.

Postacioğlu, N., Kapadia, P. D., and Dowden, J. M. The thermal stress generated by a moving elliptical weldpool in the welding of thin metal sheets. *J. Phys. D: Appl. Phys.* 30.16, 2304-12, 1997.

Postacioğlu, N., Kapadia, P. D., and Dowden, J. M. Theory of the oscillations of an ellipsoidal weld pool in laser welding. *J. Phys. D: Appl. Phys.* 24, 1288-1292, 1991b.

Postacioğlu, N., Kapadia, P. D., Davis, M., and Dowden, J. M. Upwelling in the liquid region surrounding the keyhole in penetration welding with a laser. *J. Phys. D: Appl. Phys.* 20, 340-345, 1987.

Ramakrishnan, A. On an integral equation of Chandrasekhar and Münch. *Astrophys. J.*, 115, .141-144, 1952.

Ramakrishnan, A. Stochastic processes associated with random divisions of a line. *Proc. Camb. Phil. Soc.*, 49, 473-485, 1953.

Ready, J. F. *Industrial Applications of Lasers.* Academic Press, London, 1978.

Rethfeld, B., Wendelstorf, J., Klein, T., *et al.* A self-consistent model for the cathode fall region of an electric arc. *J. Phys D: Appl. Phys.*, 29, 121-128, 1996.

Risken, H. *The Fokker-Planck Equation* (second edition). Springer, London, 1989.

Rodden, W. S. O., Solana, P., Kudesia, S. S., Hand, D. P., Kapadia, P. D., Dowden, J. M., and Jones, J. C. Melt-ejection processes in single pulse Nd:YAG laser drilling. *Proc. ICALEO'99*, Laser Institute of America, Orlando, 2000, 87, C61-69.

Rogers, S. *Some Aspects of Stefan-type Problems.* Thesis for the degree of D. Phil, Oxford University, 1977.

Rosenhead, L. *Laminar Boundary Layers.* Clarendon, Oxford, 1963.

Rosenthal, D. Mathematical theory of heat distribution during welding and cutting. *Welding J.* 20.5 220s-34s, 1941.

Rosenthal, D. The theory of moving sources of heat and its application to metal treatments. *Trans. ASME* 48 849-66, 1946.

Rykalin, N. N. and Uglov, A. A. Bulk vapor production by a laser beam acting on a metal. *High Temp. (USA)*, 9.3, 522-27, 1971.

Savina, M., Zhiyue Xu, Yong Wang, Pellin, M., and Keng Leong. Pulsed laser ablation of cement and concrete. *J. Laser Applics.* 11.6, 284-287, 1999.

Schalén, C. Uppsala Astr. Observ. Ann., Vol. 1, No. 19, 1945. Reported in: Born, M. and Wolf, E., *Principles of Optics*, p.663. Pergamon, London, 1975.

Schellhorn, M. and Spindler, G. Interaction of high-power laser radiation with metals in laser welding. *Proc. CLEO'87/IQEC'87*, Baltimore, 1987, 190.

Schellhorn, M. and von Bülow, H. CO laser deep penetration welding - a comparative study to CO_2 laser welding. Paper presented at GLL'94, Friedrichshafen, 1994.

Schulz W, Simon, G., and Vicanek, M. Ablation of opaque surfaces due to laser irradiation. *J. Phys. D: Appl. Phys.*, 19, L173-L177, 1986.

Schulz, W., Simon, G., Urbassek, H. M., and Decker, I. On laser fusion cutting of metals. *J. Phys. D: Appl. Phys.*, 20, 481-8, 1987.

Simon, G., Gratzke, U., and Kroos, J. Analysis of heat conduction in deep penetration welding with a time-modulated laser beam. *J. Phys. D: Appl. Phys.*, 26, 862-869, 1992.

Sitenko, A., Malnev, V. *Plasma Physics Theory*, Chapman & Hall London, 1995.

Smithells, C. J., Ed. *Metals Reference Book.* Butterworth, London, 1962.

Solana, P. and Negro, G. A study of the effect of multiple reflections on the shape of the keyhole in the laser processing of materials. *J. Phys. D: Appl. Phys.*, 30, 3216-22, 1997.

Solana, P., and Ocaña, J. L. A mathematical model for penetration laser welding as a free-boundary problem. *J. Phys. D: Appl. Phys.* 30, 1300-1313, 1997.

Solana, P., Kapadia, P. D., and Dowden, J. M. Surface depression and ablation for a translating weld pool in material processing: a mathematical model. *J. Laser Applics.*, 12.2, 63-67, 2000.

Solana, P., Kapadia, P. D., Dowden, J. M., and Marsden, P .J. An analytical model for the laser drilling of metals with absorption within the vapour, *J. Phys. D: Appl. Phys.*, 32 942-952, 1999.

Spitzer, L. *Physics of Fully Ionized Gases.* Interscience, New York, 1962.

Spitzer, L. and Härm, R. Transport phenomena in a completely ionised gas. *Phys. Rev.*, 89, 977-81, 1953.

Steen, W. M. *Laser Material Processing.* Springer, London, 1991.

Steen, W. M. and Courtney, C. Surface heat treatment of En8 steel using a 2 kW continuous-wave CO_2 laser. *Met. Technol.* 6, 456-462, 1979.

Steen, W. M., Dowden, J. M., Davis, M.P., and Kapadia, P.D. A point and line source model of laser keyhole welding. *J. Phys. D: Appl. Phys.*, 21, 1255-1260, 1988.

Stefan, J. Über die theorie der eisbildung, inbesondere über die eisbildung im polarmeere. *Ann. Phys. u. Chem,(Widermann) N.F.* 42, 269-286, 1891.

Stommel, H. *The Gulf Stream.* Cambridge University Press, London, 1965.

Stratton, J A. *Electromagnetic Theory.* McGraw-Hill, New York:, 500-11, 1941.

Swift-Hook, D. T. and Gick, A. E. F. Penetration welding with lasers. *Welding J.*, 52, 492s-99s, 1973.

Swokowski, E. W., Olinick, M., Pence, D., and Cole, J. A. *Calculus*, 6th edn. PWS Publishing Co., Boston, 1994.

Tayler, A. B. *Mathematical Models in Applied Mechanics*. Clarendon, Oxford, 1986.

Tix, C. and Simon, G. A transport theoretical model of the keyhole plasma in penetration laser welding, *J. Phys. D: Appl. Phys.*, 26, 2066-2074, 1993.

Tix, C., Gratzke, U., and Simon, G. Absorption of the laser-beam by the plasma in deep laser-beam welding of metals. *J. Appl. Phys.* 78, 6448-53, 1995.

Trappe, J., Kroos, J., Tix, C., *et al.* On the shape and location of the keyhole in penetration laser-welding. *J. Phys. D: Appl. Phys.* 27, 2152-54, 1994.

Tsuji, M., Nishitani, T., and Shimizu, M. Technical note: Three dimensional coupled thermal stress in infinite plate subjected to a moving heat source. *J. Strain Anal.* 31.3, 243-247, 1994.

Velasco, S. On the Brownian motion of a harmonically bound particle and the theory of a Wiener process. *Eur. J. Phys.* 6, 259-265, 1985.

Vicanek, M, Colla, T. J., and Simon, G. Hydrogen enrichment in laser-beam welding of aluminum. *J. Phys. D: Appl. Phys.* 27, 2284-90, 1994.

Vicanek, M. and Simon, G. Momentum and heat-transfer of an inert-gas jet to the melt in laser cutting. *J. Phys. D: Appl. Phys.* 20, 1191-96, 1987.

Vicanek, M., Simon, G., and Urbassek, H. M., *et al.* Hydrodynamical instability of melt flow in laser cutting. *J. Phys. D: Appl. Phys.* 20, 140-45, 1987.

Vicenti, W. G. and Kruger, C. H. *Introduction to Physical Gas Dynamics*. Wiley, New York, 1965.

Wang, M. C. and Uhlenbeck, C. E. On the theory of Brownian Motion II. *Per. Mod. Phys., I*, 17 Nos. 2 & 3, 323-342, 1945.

Weber, H. and Riemann, B. *Die Partiellen Differentialgleichungen der Mathematischen Physik*. Viewg, Braunschweig, 1919, Vol. 2.

Whipple, F. J. W. and Chalmers, J. A. On Wilson's theory of the collection of charge by falling drops. *Quart. J. Roy. Meteorological Soc.* 70, 103, 1944.

INDEX

9 780367 397319